JUST TRANSITION SOCIOECONOMIC IMPACT ASSESSMENT FOR GEORGIA

MARCH 2025

ASIAN DEVELOPMENT BANK

 Creative Commons Attribution 3.0 IGO license (CC BY 3.0 IGO)

© 2025 Asian Development Bank
6 ADB Avenue, Mandaluyong City, 1550 Metro Manila, Philippines
Tel +63 2 8632 4444; Fax +63 2 8636 2444
www.adb.org

Some rights reserved. Published in 2025.

ISBN 978-92-9277-222-2 (print); 978-92-9277-223-9 (PDF); 978-92-9277-224-6 (ebook)
Publication Stock No. TCS250081-2
DOI: http://dx.doi.org/10.22617/TCS250081-2

The views expressed in this publication are those of the authors and do not necessarily reflect the views and policies of the Asian Development Bank (ADB) or its Board of Governors or the governments they represent.

ADB does not guarantee the accuracy of the data included in this publication and accepts no responsibility for any consequence of their use. The mention of specific companies or products of manufacturers does not imply that they are endorsed or recommended by ADB in preference to others of a similar nature that are not mentioned.

By making any designation of or reference to a particular territory or geographic area in this document, ADB does not intend to make any judgments as to the legal or other status of any territory or area.

This publication is available under the Creative Commons Attribution 3.0 IGO license (CC BY 3.0 IGO) https://creativecommons.org/licenses/by/3.0/igo/. By using the content of this publication, you agree to be bound by the terms of this license. For attribution, translations, adaptations, and permissions, please read the provisions and terms of use at https://www.adb.org/terms-use#openaccess.

This CC license does not apply to non-ADB copyright materials in this publication. If the material is attributed to another source, please contact the copyright owner or publisher of that source for permission to reproduce it. ADB cannot be held liable for any claims that arise as a result of your use of the material.

Please contact pubsmarketing@adb.org if you have questions or comments with respect to content, or if you wish to obtain copyright permission for your intended use that does not fall within these terms, or for permission to use the ADB logo.

Corrigenda to ADB publications may be found at http://www.adb.org/publications/corrigenda.

Note:
In this publication, "$" refers to United States dollars and "GEL" refers to lari.

Cover design by Joe Mark Ganaban.

Contents

Tables, Figures, and Boxes

Figures

Boxes

Acknowledgments

This report was prepared with support from the Asian Development Bank (ADB) technical assistance project, TA 6811: Economic Diagnostic Studies in Asia and the Pacific (Subproject 1). It aims to help the Government of Georgia assess the socioeconomic impact of its planned climate action, which is critical for the transition to a low-carbon economy.

A team of ADB staff and consultants worked together to produce the report. Lindsay Marie Renaud led the study team. Lei Lei Song, director, Economic Analysis and Operational Support Division, Economic Research and Development Impact Department, provided overall guidance in the implementation of the study. Maria Pia Ancora, Maria Rowena Cham, Kathleen Anne C. Coballes, Swati Mitchelle Dsouza, Kate Hughes, Navina Sanchez Ibrahim, Jairus Carmela C. Josol, George Luarsabishvili, Malte Maass, Ekaterine Mikadze, Kiyoshi Taniguchi, and Maia Tserediani provided comments. Special thanks go to the staff of the Georgia Resident Mission, the peer reviewers of the Central and West Asia Department and other ADB departments who provided information, valuable suggestions and advice, facilitated field missions, and participated constructively in reviewing and commenting on the study.

The report was copyedited by Tuesday Soriano. Joe Mark Ganaban did the graphic design, layout, and typesetting. Maia Tserediani and Hannah Althea Estipona provided administrative support, and Bhabhani Dewan helped with the publication process.

The team would like to thank the Government of Georgia and its agencies, including the Ministry of Environmental Protection and Agriculture and the Ministry of Economy and Sustainable Development, for their helpful contributions.

Abbreviations

ADB	Asian Development Bank
CSAP	2030 National Climate Strategy and 2021–2023 Action Plan
CSO	civil society organization
GHG	greenhouse gas
MDB	multilateral development bank
MOESD	Ministry of Economy and Sustainable Development
MSMEs	micro, small and medium-sized enterprises
O&M	operation and maintenance
RIA	Regulatory Impact Assessment
VAT	value-added tax

Executive Summary

Georgia's climate action strategy, as detailed in its updated nationally determined contribution, the 2030 National Climate Strategy and 2021–2023 Action Plan, and the Long-Term Low Emission Development Strategy, aims to significantly reform the energy and transport sectors to achieve low-emission development. By 2050, Georgia plans to transition to a green economy with a focus on energy-efficient technologies and renewable energy. The goal is to reduce emissions from these sectors by 15% by 2030 by developing renewable energy facilities, improving energy efficiency in construction, and modernizing the transport sector.

Ensuring a just transition is essential for the effective implementation of Georgia's climate strategy, as the planned actions offer prospects for economic growth but may also lead to negative social and economic consequences. This assessment aims to assist the Government of Georgia in evaluating the socioeconomic impacts of these climate measures in the energy and transport sectors. It provides actionable recommendations to mitigate adverse impacts and enhance positive effects, identifying areas where support from the Asian Development Bank could be beneficial.

The following activities were agreed with the Government of Georgia for the assessment as part of the initial scoping of the project:

- Activity 1: Construction of renewable energy power plants (wind, solar, hydro) for a total installed capacity of 1,269 megawatts
- Activity 2: Energy efficiency reform for the construction of new buildings
- Activity 3: Intercity bus passenger transport reform

1. Socioeconomic Impact Assessment Outcomes and Findings and Recommendations

This second main part examines the results of the qualitative assessment conducted for each of the selected activities. It also presents actionable recommendations aimed at mitigating the negative impacts and maximizing the positive impacts.

Activity 1: Construction of renewable energy power plants

To meet the substantial increase in energy demand, the government, in cooperation with the private sector, is planning to build a number of renewable energy power plants with a total installed capacity of approximately 1.2 gigawatts.

The assessment highlights that the most significant challenge in implementing renewable energy projects in Georgia is their impact on local communities. This challenge is reflected in the estimated potential impact on property values, which are largely affected by negative public perceptions of renewable energy. This skepticism, stemming from past experiences with hydropower projects, could jeopardize the success of future renewable energy projects if not properly addressed.

The assessment has also shown that while the introduction of new renewable energy sources may lead to a slight increase in electricity prices, this is unlikely to significantly affect the purchasing power of most households. However, for the most economically vulnerable groups (approximately 76,000 households), even a small increase in expenditure may pose a challenge. It suggests integrating financial support for these groups within existing aid mechanisms, which requires about GEL76,000 per month.

Furthermore, the establishment of new renewable energy plants is expected to create about 4,000 job opportunities, which is particularly beneficial for regions with high unemployment. These projects also offer public administrations the opportunity to generate additional revenue and potentially fund just transition measures.

To mitigate these negative impacts and maximize opportunities, the assessment recommends the following:

(i)   Refine the regulatory framework to address environmental concerns, land use and access rights issues, and economic impacts (property value and others), especially for hydropower plants.

(ii)   Enhance positive impacts through incentives for project developers that employ local workers, investment in social infrastructure, skills development, benefit sharing with local communities, and effective tax collection.

(iii)   Develop gender programs to increase the proportion of women in energy sector jobs.

(iv)   Establish a just transition fund or provide guidelines for the use of fiscal revenues for just transition measures.

(v)   Improve communication with civil society organizations to improve public perception of renewable energy.

Activity 2: Energy efficiency reform for the construction of new buildings

The Georgian Parliament has adopted legislation on energy efficiency and energy efficiency of buildings that brings the country closer to European Union standards. The legislation aims to reduce emissions and pollution, improve the energy efficiency of buildings, decrease energy imports, and bolster the country's energy security. The reform stipulates that all construction permits issued in Georgia from July 2023 must meet a number of minimum energy efficiency requirements set out in the Law on Energy Efficiency. These include requirements for the energy performance of buildings, building parts or elements, as well as requirements for the energy performance of a building's engineering and technical support systems.[*]

The adoption of the energy efficiency reform for the construction of new buildings presents a mixed bag of opportunities and risks. It has the potential to promote economic development and social welfare but also poses challenges for certain segments of the population. Upstream, the main impact is the risk to micro, small and medium-sized enterprises as suppliers of construction materials. This is primarily due to the potential need for these businesses to invest in new machinery and equipment to comply with the new standards. While this challenge may seem daunting at first, it also presents an opportunity to boost the growth of an underdeveloped sector in the country. Tailored financing programs, capacity-building initiatives, and stimulation of local demand should be implemented to overcome this challenge.

Downstream, the most significant impact is the potential savings in energy consumption, especially for the lowest-income households who spend a higher proportion of their income on this expenditure. On the other hand, contrary to initial concerns, the impact on the rental and acquisition of houses is estimated to be minor or negligible due to the socioeconomic conditions in Georgia (high ownership rates and safe debt ratios).

However, these positive impacts depend on access to energy-efficient buildings. The main challenge is therefore to ensure that low-income households have access to these new energy-efficient properties so that they too can benefit from the potential savings in energy consumption.

The recommended measures initially focus on ensuring that the reform is socially accepted and well received by the market. To achieve this, it is recommended to focus on awareness-raising campaigns to showcase the potential benefits of the reform to the different segments of the population.

Once the market is established, the authorities should promote the implementation of energy efficiency programs for low-income households by offering financial incentives to low- and middle-income households to access this type of property. This could include fiscal benefits, tax breaks, or subsidies for households that purchase or rent energy-efficient buildings. Following the example of some European countries, Georgia can take measures to provide fiscal benefits and advantages to households that acquire new energy-efficient buildings and agree to rent them at an affordable price for low- and middle-income households. These types of measures aim to provide benefits to both buyers and renters and expand access to energy-efficient housing.

As Georgia goes through this transformation phase, the success of the reform lies not only in mitigating the negative impacts, but also in the equitable distribution of its benefits. The government's commitment to implementing these recommendations will determine the degree of inclusivity achieved and set the stage for a sustainable and just transition.

---

[*]   Parliament of Georgia. 2020. Law of Georgia on Energy Efficiency of Buildings. Legislative Herald of Georgia. https://www.matsne.gov.ge/en/document/view/4873932?publication=0.

Activity 3: Intercity bus passenger transport reform

Aiming to reduce greenhouse gas emissions in the transport sector, the government is planning a reform of intercity passenger transport that will create a new legal framework for the sector. The aim of the reform is to improve the quality of intercity passenger transportation services by creating a competitive environment and improving safety levels and environmental aspects. A key focus is on addressing the disorderly bus traffic, which originates from crowded gathering places instead of designated bus stations, leading to increasing traffic congestion and environmental degradation. The reform aims to curb the practices of unregistered carriers that operate without schedules and violate traffic regulations, environmental norms, and drivers' working hours.

The assessment aims to measure the impact, focusing on the population groups that are more exposed and vulnerable to ticket price oscillations, job market changes, and modifications in service characteristics.

The intercity transport reform in Georgia presents both challenges and opportunities for the various stakeholders. The assessment identifies the potential increase in ticket prices as a significant impact, particularly for low- and middle-income groups that rely on this service for essential activities. This could affect their disposable income, although the impact could be mitigated through cash transfer programs and improvements to safety and reliability. Upper-middle and high-income groups that do not currently rely on the service could be drawn to use it through service improvements and communication campaigns.

For service providers, the financial and administrative challenges are small but need to be addressed. The main costs are the purchase of new vehicles, estimated at $39,000 for fleet renewal (per vehicle). Support measures such as preferential financing and de-risking programs as well as training and support centers can help informal drivers in particular to adapt to the new requirements.

The reform also offers the opportunity to increase women's participation in the labor market, improve safety, and reduce harassment on public transport. However, it may have an impact on workers who rely on informal jobs at existing bus stations. Local authorities should take measures to support these workers, such as dedicated spaces for local artisans and community development initiatives.

Connectivity to remote villages is crucial. This can be achieved through careful planning and alternative transportation measures. Regarding gender issues, the reform offers an opportunity to address the disproportionate economic burden of transportation costs on women. Policies should ensure women's financial access to intercity transportation in addition to measures to improve safety, punctuality, and reliability. Continued monitoring of these measures is critical to assess their effectiveness and ensure compliance. If these gender-related issues are addressed, the reformed public transportation system will not only provide a safer, more efficient, and reliable mobility solution for women but will also contribute to a more just and equitable society for all Georgians.

2.   Conclusion

While the assessment has indeed revealed potential negative socioeconomic impacts that could hinder the implementation of climate measures in Georgia's energy and transport sectors, it has also unveiled significant opportunities. Successfully mitigating adverse impacts and leveraging these opportunities is crucial for a just transition.

The study provides a list of recommended actions that could be implemented by national authorities based on the identified impacts and opportunities and the significance of each of them. Implementing these recommendations will require considerable planning and strategic efforts, such as managing economic disruption, raising public awareness, adopting a comprehensive structural approach, and consulting with local communities and stakeholders. The insights gained in this report will guide informed decision-making but implementing these plans could be a complex and significant undertaking.

It is also important to recognize that the results of this study provide a picture at a sectoral level. For more detailed and nuanced insights, a project-level social and environmental assessment could be useful. This complementary approach would provide granular information that would enable a deeper understanding of the specific impacts and opportunities within individual projects.

Overview

This report summarizes the key findings and recommendations of the Just Transition Socioeconomic Impact Assessment for three selected activities that the Government of Georgia will implement—or is already implementing—in line with the country's climate strategy.

The report is divided into three main sections: (i) the introduction sets out Georgia's climate commitments and actions, the objective and scope of the assessment, including a description of the criteria for selecting the activities to be assessed, and an overview of the related initiatives; (ii) the assessment and recommendations on the selected activities; and (iii) overall conclusions.

CHAPTER I

Introduction

1. Georgia's Climate Action in Context

Georgia's climate commitments are outlined in its updated nationally determined contribution, the 2030 National Climate Strategy and 2021–2023 Action Plan (CSAP), and the Long-Term Low Emission Development Strategy, which highlight the energy and transport sectors as key sectors for the transformation toward Georgia's long-term vision of low-emission development.

In 2017, the energy sector accounted for 62.4% of the country's total greenhouse gas (GHG) emissions. In terms of energy consumption, the transport subsector accounted for the largest share of the energy sector (41%, or 3,901 gigagrams of carbon dioxide equivalent), followed by the residential sector (21%, or 2,008 gigagrams of carbon dioxide equivalent).[1]

Georgia intends to go "green" by 2050 by switching to energy-efficient technologies and renewable energy. Because of the significance of the energy and transport sectors, Georgia has committed to transform these sectors and reduce emissions in each of them by 15% by 2030 compared to the baseline (2015) (footnote 1). Sectoral priorities in the energy sector include the development of large-scale renewable energy production facilities, the implementation of "large-scale energy-efficient measures" in the construction sector, and the development of "energy-efficient technologies" in existing industries (footnote 1). In the transport sector, the Government of Georgia plans to modernize infrastructure and harmonize national legislation to meet international standards.

2. Just Transition: A Key Factor in the Success of Climate Action

First, it is important to define the meaning of the term "just transition." Although there is no universally accepted definition, a commonly accepted one is that of the International Labour Organization.[2]

---

[1] Government of Georgia. 2023. Georgia's Long-Term Low Emission Development Strategy.
[2] International Labour Organization. https://www.ilo.org/.

A Just Transition means greening the economy in a way that is as fair and inclusive as possible to everyone concerned, creating decent work opportunities and leaving no one behind. A Just Transition involves maximizing the social and economic opportunities of climate action, while minimizing and carefully managing any challenges—including through effective social dialogue among all groups impacted, and respect for fundamental labor principles and rights. Ensuring a just transition (...) is important for all economic sectors—by no means limited to energy supply—and in urban and rural areas alike.

Another key reference point for this assessment was the Multilateral Development Bank (MDB) Group's Just Transition High Level Principles (Figure 1), ensuring that the study is aligned with the Asian Development Bank's (ADB) approach to supporting a just transition.[3]

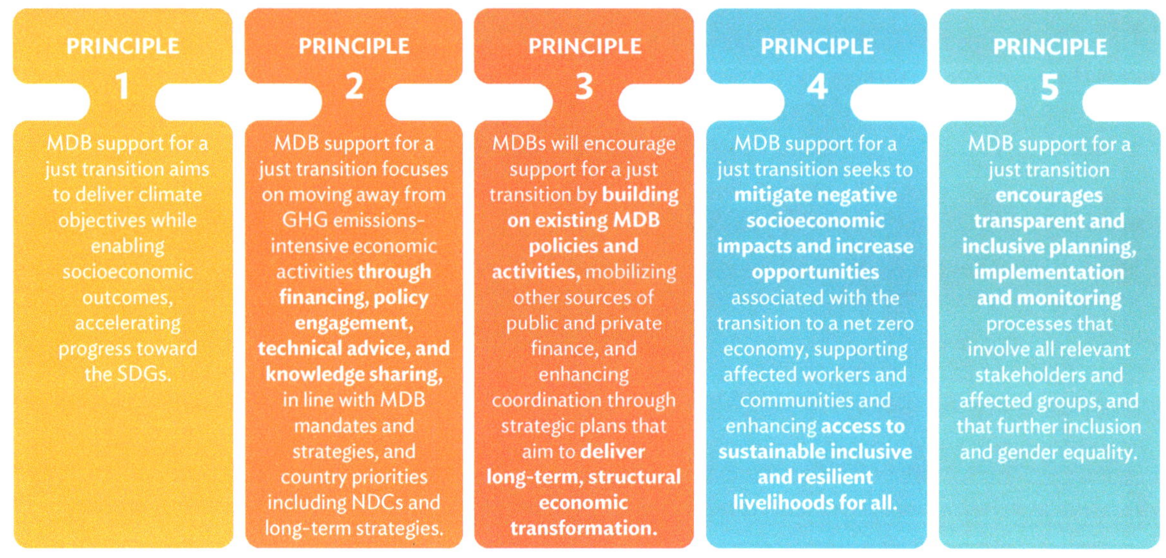

Figure 1: Just Transition High-Level Principles of the Multilateral Development Bank Group

GHG = greenhouse gas, MDB = multilateral development bank, NDC = nationally determined contribution, SDGs = Sustainable Development Goals.
Source: MDB Group. n.d. MDB Just Transition High-Level Principles. https://www.adb.org/sites/default/files/related/238191/MDBs-Just-Transition-High-Level-Principles-Statement.pdf.

In line with these two references, ensuring a just transition is essential for the effective implementation of Georgia's climate strategy. While the planned climate actions offer prospects for fair and inclusive growth—particularly with supportive measures and policies—they also come with challenges, costs, and risks that require careful management and mitigation for a just transition for all.

---

[3]   MDB Group. n.d. MDB Just Transition High-Level Principles. https://www.adb.org/sites/default/files/related/238191/MDBs-Just-Transition-High-Level-Principles-Statement.pdf.

3.   Objective of the Assessment

The main objective of this assessment is to support the Government of Georgia in assessing the potential socioeconomic implications of the climate measures planned by the government. These measures specifically target the country's two major emitting sectors—energy and transport—as part of the broader initiative to transition to a low-carbon economy.

The outcomes of the impact assessment serve as the basis for a series of recommendations proposing measures that the government can subsequently adopt. These recommendations are designed to mitigate the negative impacts and enhance the positive impacts, including areas where the government can benefit from ADB support.

The study quantifies the main relevant impacts of the activities defined in each sector and assesses in detail the impacts deemed material. It assesses the extent to which different social groups will be affected by the changes in the energy and transport services. The assessment also examines the repercussions along the value chain of the sectors directly affected by Georgia's mitigation actions. It analyzes the impacts on various aspects, including employment, government revenues, costs, local communities, vulnerable social groups, and new market dynamics.

The assessment was conducted using the best available data sources, with preference given to primary data whenever possible. In cases where primary data were not available, hypotheses, averages, assumptions, and estimates from international literature were used.

The value of the study lies in its ability to guide government decision-making. The figures obtained in the calculations, per se, are not the main goal of this assessment. Rather, what is most important is the extent to which they empower the government to make informed decisions about the measures that could be taken to mitigate the negative impacts and enhance the positive ones.

4.   Scope of the Assessment

(a)   Activities Selected for the Assessment

The selection of activities was not predetermined but was deliberately included in the scope of the assessment to ensure alignment with government priorities and needs. The mitigation activities to be assessed in this study were selected in collaboration with government counterparts and ADB staff.[4] This selection process carefully considered the actions and priorities articulated in the country's nationally determined contribution, CSAP, Long-Term Low Emission Development Strategy, and other relevant documentation.

---

[4]   For transport activities with representatives of the Transport and Logistics Development Policy Department at the Ministry of Economy and Sustainable Development (MOESD). For energy generation and energy efficiency-related activities with representatives from the Energy Efficiency and Renewable Energy Department in MOESD. Details are provided in section Stakeholder Engagement.

Activities were selected through a qualitative assessment of the following:

- Potential to create socioeconomic impacts (negative and/or positive)
- Degree of social acceptance
- Potential for mitigation measures
- Suitability for modeling
- National priority

Based on these criteria, the final selected activities for each sector are presented below.

Energy Sector

Activity 1: Construction of renewable energy power plants (wind, solar, hydro) for a total installed capacity of 1,269 megawatts

To address the substantial rise in energy demand, the government is planning to build a number of renewable energy power plants in cooperation with the private sector. The list of power plants was provided by the government counterparts and can be found in Chapter II.

The impact assessment is conducted at the sectoral level, except for the impacts on local communities, which include some characteristics at the regional level. Detailed impact assessments for each individual project shall eventually be done in accordance with legal requirements and lenders' safeguard requirements and are not part of this study.

Activity 2: Energy efficiency reform for the construction of new buildings

The Georgian Parliament has adopted legislation on energy efficiency[5] **and energy performance of buildings, bringing the country closer to European Union (EU) standards.** The legislation aims to reduce emissions and pollution, improve the energy efficiency of buildings, reduce energy imports, and bolster the country's energy security. Energy demand in buildings is expected to increase as a result of gross domestic product growth in the coming decade, increasing direct and indirect emissions in this part of the sector by 150% by 2030.[6]

One of the main objectives stated in the CSAP is to support development of low-carbon approaches in the building sector by promoting climate-smart and energy-efficient technologies and services (footnote 6). This reform stipulates that all building permits issued in Georgia from July 2023 will have to meet a set of minimum energy efficiency requirements defined by the national authorities. Project submissions must include declarations confirming that the construction projects meet the new requirements. The Ministry of Economy offers training courses on the declarations for architects and project designers.

---

[5]   Agenda.GE. 2020. Georgian Parliament Adopts Energy Efficiency Legislation.
[6]   Government of Georgia. 2021. *Georgia's 2030 Climate Change Strategy.* https://mepa.gov.ge/En/Files/ViewFile/50123.

Transport Sector

One of the key objectives for reducing GHG emissions in the transport sector is to promote the use of public transport (buses, metros, and minibuses) over private cars, which are associated with the high intensity of GHG emissions in this sector.

To achieve this goal, the government is planning to implement a reform of intercity passenger transport reform to create a new legal framework for the sector. This will address the main problems associated with the level of informality and lack of compliance with existing safety and technical standards, the geographic concentration of bus stations, the old station infrastructure, the outdated bus fleet, route congestion, informal pricing, and timetables.

A core element of the reform is the compliance of existing bus stations and bus operators with the proposed package of policy amendments. The reforms aim to require all bus stations and bus operators—a large proportion of which are run informally by *kishniks* (individuals without registration), according to the Regulatory Impact Assessment (RIA)—to obtain a license by obligating them to meet a number of financial and technical requirements, including paying the necessary fees, proving the possession of a certain amount of funds, and operating vehicles that meet certain technical standards. The reform also seeks to streamline existing routes.

(b)   Impacts Considered

Table 1 lists the types of impact considered for each of the three activities:

Table 1: Types of Impacts Considered in the Assessment

Type of Impact	Definition
Direct	Direct impacts refer to the immediate and primary effects resulting from a specific action, event, policy, or decision.
Indirect	Indirect impacts refer to the secondary or unintended effects resulting from a particular action, policy, or event. In the case of this study, indirect impacts refer to those affecting the value chain[a] functioning of the sector being studied upstream and downstream, the national and local authorities, and the local communities indirectly affected by the activity.
Induced	Induced impacts refer to changes that the studied activity will trigger in aggregated socioeconomic variables, the public budget, and communities not directly targeted by the policy as a result of the direct and/or indirect impacts.

[a]   A "value chain" is defined as encompassing all activities upstream and downstream of the core activity—including the supply chain.
Source: Author's own elaboration.

(c)   Exclusions of the Assessment

The following elements are not covered in the scope of this assessment:

- This study does not quantify or qualify the potential positive impacts on air quality, water pollution, and other environmental issues, nor the potential positive impacts on the health of specific population groups.
- The assessment is made at the sector/activity level. Impact assessments at the project level are not covered.
- The assessment outputs are meant to provide an order of magnitude of potential impacts and opportunities. It should be noted that the results may vary depending on local conditions and specificities at project level.
- The assessment is primarily based on a static model. It should be noted that the results may vary depending on changes in consumption patterns, technology, inflation, etc.
- Some impacts were not included in the assessment as they were not considered relevant or significant enough or suitable for quantitative analysis.

5.   Methodology

The assessment was carried out in two main phases, as shown in Figure 2.

Figure 2: Flow of Assessment

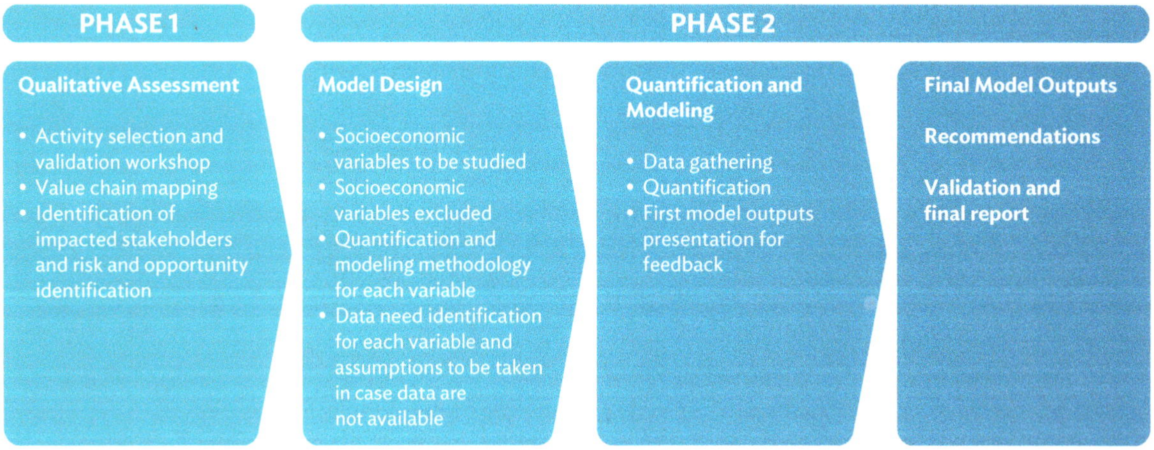

Source: Author's own elaboration.

The methodology followed to perform each task is presented below.

(a)   Phase 1: Qualitative Assessment

Value Chain Mapping

In this first step, the sequence of main activities and processes for implementing government initiatives was presented. The value chain for each activity was illustrated to identify participating stakeholders at each stage—upstream and downstream—and the types of impacts to which they are potentially exposed.

Identification of Affected Stakeholders and Risks and Opportunities

Building on the value chain mapping, the stakeholders affected by the initiatives were identified. These include companies supplying products or services, workers contributing to project development, stakeholders involved in the new market dynamics, impacts on the government budget, and participants from affected communities. Both risks and opportunities are assessed in the analysis.

Outline of Data Needs, Assumptions, and Data Availability at Pre-Assessment

As part of the qualitative assessment, a preliminary outline of data needs and possible assumptions in the event that primary data are not readily available was prepared.
This preliminary outline was evaluated by the International School of Economics at Tbilisi State University to provide an initial estimation of data availability based on its experience and knowledge. The purpose of this pre-assessment is to determine the likelihood of relevant data being available for model development. More detailed data requirements are specified in Phase 2.

(b)   Phase 2: Quantitative Assessment and Modeling

The quantitative modeling process involves several critical steps, which are described below.

Model Design Process

Based on the pre-assessment of data availability in Phase I, the most appropriate model for quantifying each of the socioeconomic variables included in the assessment was designed.
For each socioeconomic variable, a technical card was elaborated containing the following:

(i)   Description of the variable and relevance,

(ii)   Quantification of the formula and scenarios used,

(iii)   Data needs and data sources, and

(iv)   Data assumptions if data are not available.

The technical cards of the individual socioeconomic variables can be found in Appendix 1: Socioeconomic Variables Technical Cards.

Data Gathering

Data gathering was carried out in collaboration with the International School of Economics at Tbilisi State University and other project partners to meet the specific requirements of the model data. The quantification and modeling process involves performing calculations using the collected data for each target activity, following the model design indications and using the best available data.

Quantification and Modeling Process

Calculations were performed using the data collected for each of the targeted activities, following the model design specifications and using the best available data.

Impact Significance Assessment

Impact significance is the product of impact severity and impact likelihood. The significance assessment is key to determine measures that are aligned with the magnitude of the impact (Figure 3).

Figure 3: Impact Significance Assessment Matrix

		Impact Likelihood				
		Extremely Unlikely (1)	Unlikely (2)	Neutral (3)	Likely (4)	High Likelihood/ Inevitable (5)
Severity	**Low** (1)	Negligible	Negligible	Negligible	Minor	Minor
	Medium (3)	Negligible	Minor	Minor	Moderate	Moderate
	High (5)	Minor	Moderate	Moderate	Major	Major
	Critical (9)	Minor	Moderate	Major	Major	Critical

Source: Author's own elaboration.

The following criteria are used to qualitatively determine each score:

- **Severity.** The degree of scale, intensity, or importance of the impact/opportunity, categorized as low, medium, high, or critical. Table 2 shows the severity levels for the different criteria.

Table 2: Severity Assessment Criteria

Severity	Criteria
Low	Negligible or low impact/opportunity for a specific small segment
Medium	Moderate impact/opportunity that affects or benefits a specific small segment or low impact on a large group of the population
High	Significant impact/opportunity with considerable consequences or benefits to a noticeable segment
Critical	Severe impacts/opportunities that profoundly affect a wide range of stakeholders

Source: Authors' own elaboration.

- **Likelihood.** The probability of the event occurring, classified as extremely unlikely, unlikely, neutral, likely, highly likely/inevitable. Table 3 shows the likelihood levels for the different criteria.

Table 3: Likelihood Assessment Criteria

Likelihood	Criteria
Extremely unlikely	Almost impossible to occur
Unlikely	Low probability of occurrence
Neutral	Neither unlikely nor likely; 50–50 chance
Likely	Moderate to high probability of occurrence
Highly likely/Inevitable	Almost certain to occur

Source: Authors' own elaboration.

Presentation and Validation of the Model Outputs

The first results of the model were presented to government counterparts and civil society organizations (CSOs) in an in-country workshop to obtain feedback and/or validate the results. This feedback informed the elaboration of final model outputs, which were then incorporated into a comprehensive report along with recommendations.

Recommendations

The final recommendations that emerged from the conclusive outputs included a set of measures to help authorities address potential negative impacts and capitalize on opportunities arising from the implementation of the activities.

Final Presentation

The final model outputs and the recommendations were presented in a final in-country workshop attended by government counterparts and CSOs.

(c)   Stakeholder Engagement

Stakeholders involved in the project included national and local government counterparts, as well as representatives from civil society groups and organizations, employers, industry associations, academia, communities, and others (Table 4).

Table 4: List of Stakeholders

Activity	Internal Stakeholders (Government Officials)	External Stakeholders
Transversal to all activities	Kakha Lomashvili – Ministry of Environmental Protection and Agriculture of Georgia Salome Oboadze – donor relations focal	Erekle Shubitidze, ISET Gender Equality Working Group
Energy efficiency	Margalita Arabidze, Head of Energy Efficiency and Renewable Energy Department, MOESD Omar Tsereteli, Deputy Head of Energy Efficiency and Renewable Energy Department, MOESD David Advadze, Head of Sustainable Development Promotion Division of Energy Efficiency and Renewable Energy Department, MOESD Rusudan Sulamanidze, Deputy Head of Construction Policy Department	George Abulashvili, Director of Energy Efficiency Center of Georgia Karina Melikidze – SDAP Vakhang Begashvili, GOPA-Intec Levan Natadze, Green Building Council Georgia Ana Sabakhtarishvili, CEO, Infrastructure Construction Companies' Association Levan Vepkhvadze, Business Association of Georgia
Renewable energy	Jubo Turashvili – Head of Energy Policy and Investment Projects Department, MOESD	Nino Purtskhvanidze – Ministry of Regional Development and Infrastructure Kakha Khandoshvili – Ministry of Education Marine Gogoladze – National Statistics Office of Georgia Ramaz Oboladze – Market Surveillance Agency Temur Bolotashvili – Tbilisi City Hall – Head of Architecture
Transport	Tsabuna Nakulashvili – Transport and Logistics Development Policy Department, MOESD	Association of Transport and Roads of Georgia Partnership for Road Safety Bus Station NIGE LLC Georgian Bus Land Transport Agency

GOPA = International Energy Consultants, ISET = International School of Economics at Tbilisi State University, MOESD = Ministry of Economy and Sustainable Development, NIGE = Didube Bus Station Nige (this is a company providing transport service), SDAP = Sustainable Development and Policy.
Source: Author's own elaboration of stakeholder consultations.

Stakeholder engagement was a priority throughout the project. Each step of the project was presented by the project team and validated by relevant government counterparts (Table 5).

Table 5: Stakeholder Consultations Throughout the Project

	Stakeholder Consultation	Description/Objective/Output	Stakeholders Consulted/Engaged
1	Activity definition workshop (remote)	Define the list of activities to be included in the scope of the assessment.	Government counterparts and ADB staff for all the relevant activities
2	Qualitative assessment and model design presentation (remote)	Present the qualitative assessment, model design, impacts included or excluded, data needs, and validate the qualitative assessment.	Government counterparts
3	Presentation of quantification assessment (in-country mission)	Present the model results and first conclusions, explore alternatives for missing data, and validate the results.	Government counterparts and CSOs from different sectors (such as energy, transport, and environment)
4	Final report for review and validation	Present the final report for review and gather feedback.	ADB, government counterparts
5	Final presentation (in-country mission)	Present the final report for review, gather feedback, and validate results.	Government counterparts and CSOs from different sectors (such as energy, transport, environment)

ADB = Asian Development Bank, CSO = civil society organization.
Source: Author's own elaboration.

CHAPTER II

Socioeconomic Impact Assessment Findings and Recommendations

This second main section delves into the outcomes of the qualitative assessment carried out for each of the selected activities. It also presents actionable recommendations aimed at mitigating the negative impacts and maximizing the positive impacts.

Activity 1: Construction of the Renewable Energy Power Plants

1.   Description of Activity

List of Renewable Energy Plants and Construction Scenario

To meet the increasing demand for electricity, Georgia intends to promote the development of solar, hydro, and wind renewable energy power plants as part of its climate action plan.

The list of renewable energy power plants to be included in the assessment, as well as the year in which each power plant is to be built, was provided by government counterparts based on their national priorities (Tables 6, 7, and 8).

The increased installed capacity is not intended to replace existing fossil fuel capacity or reduce energy imports. It is intended to meet the increase in energy demand and limit the growth of energy imports.

Table 6: Solar Power Plants in Georgia

Solar Power Plants		
New Renewable Energy Power Plant	**Capacity** (MW)	**Year**
Udabno (Kakheti)	6.08	2026
Iliatsminda (Kakheti)	50	2026
Geosteel (Kvemo Kartli)	30	2026
Marneuli (Kvemo Kartli)	68	2025
Akhaltsikhe (Samtskhe-Javakheti)	10	2026
Sakuneti (Samtskhe-Javakheti)	8	2026
Total new solar power capacity	172.08	

MW = megawatt.
Source: Government of Georgia. 2021. Georgia's 2030 Climate Change Strategy.

Table 7 is a list of all the wind power plants in the country and their respective capacities.

Table 7: Wind Power Plants in Georgia

Wind Power Plants		
New Renewable Energy Power Plant	**Capacity** (MW)	**Year**
Imereti 1 (Imereti)	100	2026
Rikoti (Shida Kartli)	20	2025
Pirveli (Shida Kartli)	110	2027
Tbilisi	54	2027
Zeda Tseva (Imereti)	50	2025
Nigoza (Shida Kartli)	54	2027
Kaspi (Shida Kartli)	8	2024
Ruisi (Shida Kartli)	50	2027
Samgori (Kvemo Kartli)	49	2027
Chero Energy (Shida Kartli)	28	2026
Kareli (Shida Kartli)	206	2028
Total new wind power capacity	729	

MW = megawatt.
Source: Government of Georgia. 2021. Georgia's 2030 Climate Change Strategy.

Table 8 is a list of all the hydropower plants in the country and their respective capacities.

Table 8: Hydropower Plants in Georgia

Hydropower Plants		
New Renewable Energy Power Plant	**Capacity** (MW)	**Year**
Kirnati (Adjara)	51.25	2023
Khobi 2 (Samegrelo-Zemo Svaneti)	46.70	2023
Mtkvari (Samtskhe-Javakheti)	54.10	2024
Chiora (Racha-Lechkhumi-Kvemo Svaneti)	33.60	2025
Stori 1 (Kakheti)	26.28	2026
Samkuristskali 2 (Kakheti)	26.10	2023
Metekhi 1 (Kvemo Kartli)	36.70	2027
Ghebi (Racha-Lechkhumi-Kvemo Svaneti)	14.34	2027
Lukhuni 2 (Racha-Lechkhumi-Kvemo Svaneti)	33.60	2024
Zoti Cascade (Guria)	46.06	2026
Total new hydropower capacity	368.73	

MW = megawatt.
Source: Government of Georgia. 2021. Georgia's 2030 Climate Change Strategy.

2.   Value Chain Map

Figure 4 shows the general value chain for the construction of renewable energy power plants with a combined capacity of approximately 1 gigawatt.

Figure 4: Value Chain for the Construction of Renewable Energy Power Plants in Georgia

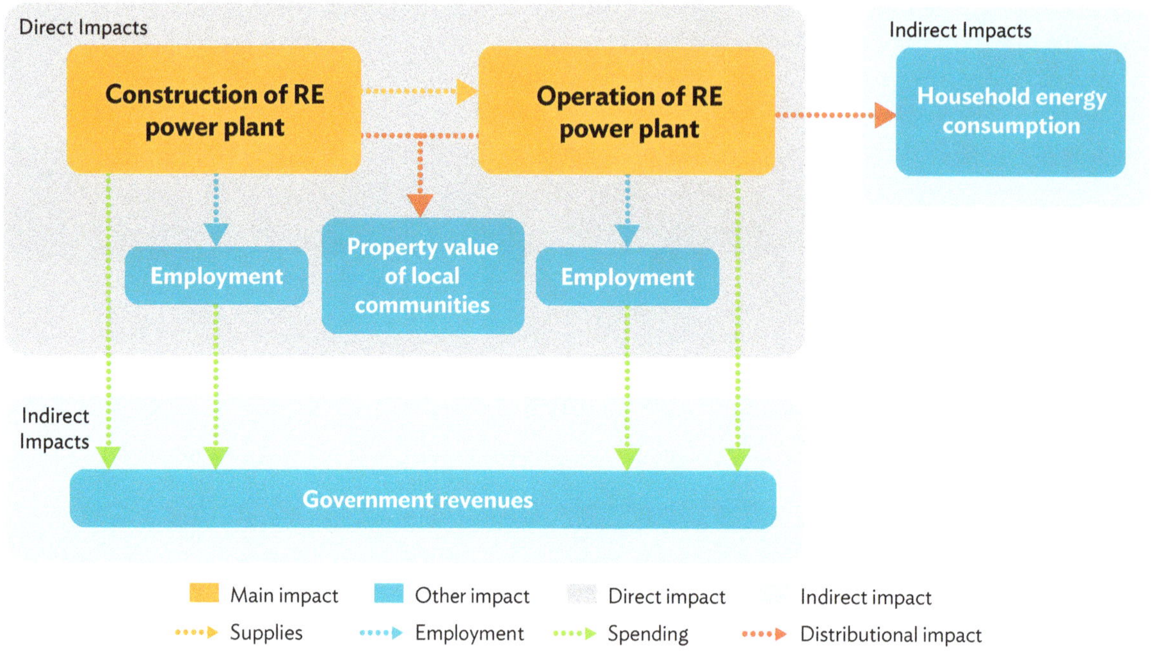

RE = renewable energy.
Source: Author's own elaboration.

3.   Qualitative Assessment: Identification of Affected Stakeholders and Risks and Opportunities

As the new renewable energy power plants are additional installed capacity (and not a replacement of existing capacity), the primary (positive) impact of this activity is the creation of new jobs both in the construction of the planned renewable energy (RE) power plants and in their subsequent operation. This positive outcome can be maximized if the project prioritizes the employment of local people.

Another important issue to consider is the impact on the local communities in the vicinity of the new RE power plant. This can have positive effects, such as the creation of new jobs for the local people, especially if the project implements a local employment policy. On the other hand, it can also have negative impacts for the local communities, which could manifest in problems such as a decrease in property values, restriction of land use rights, negative effects on agricultural businesses, negative effects on tourism, etc. There are various policy approaches that can be implemented for each type of renewable energy to maximize and/or minimize these positive or negative impacts. While this assessment does not quantitatively measure each of these aspects—as this would require a project-level assessment—it aims to provide general policy recommendations based on the literature and previous examples of similar projects.

Government revenues may increase because taxes on the income of those newly employed in both the construction and the operation of new RE power plants increase, and because of the corporate taxes paid by these entities as they carry out their operations. At the same time, the government may consider granting tax incentives to renewable energy power plants, which could reduce the amount of revenue generated by these new facilities. Further spending to subsidize renewable energy generation may further strain government resources. Moreover, the government may decide to cut subsidies for other energy sources, such as natural gas-fired thermal power plants, to prioritize incentives for renewable energy.

Households could also be affected if electricity prices rise or fall as a result of the incorporation of the planned RE power plant facilities.

The planned projects are not meant to displace natural gas-fired thermal power plants. Therefore, the impact in this sector is not considered.

There are other sectors that can be considered as part of the value chain of RE power plant construction, but whose activities are not considered significant enough in the Georgian context. These include impacts such as the transportation of construction materials for renewable energy plants. In addition, the manufacture of RE equipment is not taken into account, as these materials are imported into Georgia. The construction of the necessary transmission lines for the integration of RE into the grid is also not considered, as these projects are carried out by international companies in Georgia using international workforce.

It should also be noted that the current construction of the Black Sea submarine cable[7]—which is expected to be operational from 2029—will allow Georgia to export energy to Europe, especially in summer when national energy demand is lower. However, the additional installed capacity analyzed in this assessment (1.2 gigawatts) is not considered significant enough to generate significant electricity exports—it will mainly be used to meet increasing national electricity demand.

---

[7]   European Commission. 2022. Statement by President von der Leyen at the Signing Ceremony of the Memorandum of Understanding for the Development of the Black Sea. https://ec.europa.eu/commission/presscorner/detail/en/statement_22_7807.

4.   Socioeconomic Variables Selected for the Study

Following the qualitative assessment and discussions with government counterparts, the following socioeconomic variables were selected for the study.

In Table 9, variables 1–5 are most significant from a just transition perspective as they may have opportunities and/or negative impacts on society. Variables 6–9 were included as they can create a new source of public revenue that can be used to fund just transition measures.

Table 9: Different Socioeconomic Variables and Their Descriptions

	Socioeconomic Variable	Type of Impact	Description
1	Distributional impact on electricity expenditure in household groups	Indirect	Indicates how much the change in electricity prices affects the share of income spent on electricity by each household income group, including the impact on women-led households.
2	Job[a] creation for renewable energy construction	Direct	Number of direct full-time equivalent (FTE) jobs created during the construction of the renewable energy power plants. Also assesses the creation of employment opportunities for women.
3	Job creation for renewable energy projects operation and maintenance	Direct	Number of direct FTE jobs created for the operation and maintenance of the new renewable energy power plants. Also assesses the creation of employment opportunities for women.
4	Job opportunities for local communities vs. unemployment	Direct	Indicates the job opportunities (temporary and permanent) that will be created for local communities, taking into account the number of unemployed people at each location who may benefit from this opportunity. Also assesses the creation of employment opportunities for women.
5	Negative impact on property values of surrounding areas	Direct	Although not perfect, it is assumed that property value reflects and encompasses other environmental, economic and social issues faced by local communities as they affect the property value in one way or another.[b]
6	Government subsidies	Indirect	Government subsidies for renewable energy projects and a decrease in subsidies for gas imports.
7	Government tax revenue from new renewable energy projects	Indirect	Tax revenues are generated through taxes imposed on various aspects of the renewable energy project, such as corporate income tax, property tax, and value-added tax.
8	Government nontax revenue from renewable energy projects	Indirect	Nontax revenues may include revenues from fees and permits as well as other contributions such as royalties and lease payments for land resources used.
9	Government tax revenues from new employees	Indirect	Taxes collected from the people who would become employed thanks to these projects.

[a]   In this report a "job" is defined as a full-time equivalent employment in Georgia: an employee who works a 40-hour work week.
[b]   C. Bohlen and L. Y. Lewis. 2009. Examining the Economic Impacts of Hydropower Dams on Property Values Using GIS. *Journal of Environmental Management*. 90 (Suppl 3). pp. S258–69. doi: 10.1016/j.jenvman.2008.07.026.
Source: Author's own elaboration after stakeholder consultations.

Exclusions of the Assessment

There is no evidence that natural gas-fired thermal power plants will be displaced by renewable energy. Therefore, the impacts in this sector are not considered, apart from the possible reduction in government spending, such as subsidies, as mentioned in Table 9.

There are other sectors that can be considered as part of the value chain of renewable energy power plant construction, but whose activities were considered too small to cause significant impacts in the Georgian context. These include the transportation of construction materials for the renewable energy plants and the construction of the necessary transmission lines for the integration of renewable energy into the grid. In addition, the manufacturing of renewable energy equipment is not taken into account, as these materials are imported into Georgia. These assumptions will be reviewed again during the second phase to ensure that this is indeed the case.

5.   Summary of the Model Outputs

This subsection presents the key findings for each socioeconomic variable, including a summary of the model outputs and recommendations for government counterparts.

(a)   Variable 1: Distributional Impact on Electricity Expenditure in Household Groups

Summary of Model Outputs

Potential price increase: Analysis of the Georgian context

It is important to note that the Georgian context has various unique circumstances and dynamics that could suggest a nuanced influence on the impact of renewable energy integration on electricity prices. These include the following:

- **Electricity price regulation and open electricity market.** The price of electricity in Georgia is regulated by the Georgian National Energy and Water Supply Regulatory Commission (GNEWRC). Although the GNEWRC is formally independent, it follows the recommendations of the Georgian government. This regulatory framework has led to relatively low and constant electricity prices for households.

  The Georgian government has implemented a resolution advocating an open electricity market. This shift in dynamics is expected to have a significant impact on prices, while the introduction of new renewable energy can only mitigate this impact to a certain extent.

  As the renewable energy power plants considered in this assessment will start generating electricity from 2025, it is likely that they will be operating in an already open electricity market.

- **Current share of renewable energy in electricity generation.** Georgia has one of the highest shares of renewables in its electricity mix (81.1% in 2021).[8] Adding more renewable energy to the mix would not have the same effect as in other countries that rely on fossil fuels.

- **Price composition.** In Georgia, 34% of the total electricity price paid by users goes to the electricity producers (Figure 5). This allocation is notably lower than in other countries where electricity generation constitutes a larger share of the total price.[9] Therefore, the addition of new renewable energy power plants is likely to have a relatively smaller impact on the final price of electricity in Georgia.

Figure 5: Electricity Tariff Structure, Georgia Versus Europe

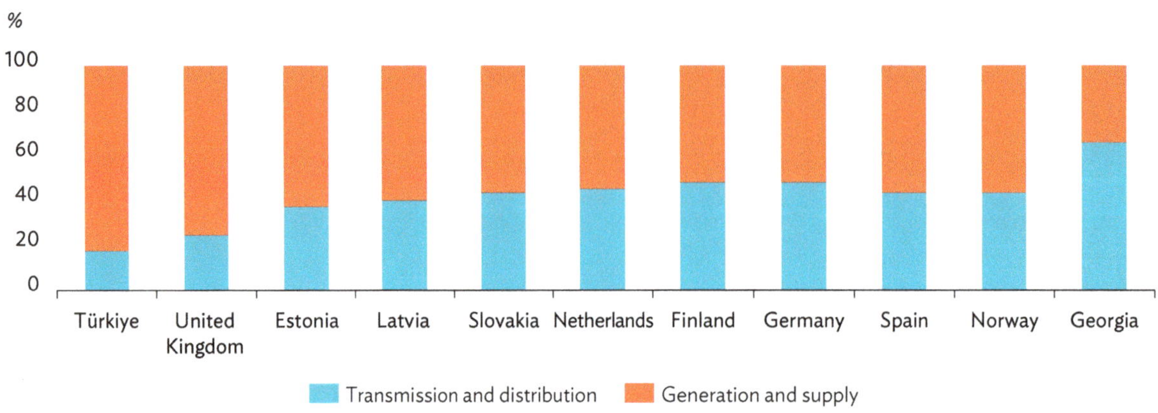

Source: E. Livny and I. Galdava. 2012. Everything You Wanted to Know About Your Electricity Bill. ISET Economist Blog. ISET Policy Institute. https://iset-pi.ge/en/blog/619-everything-you-wanted-to-know-about-your-electricity-bill.

Price increase scenario used for the analysis

Given the Georgian context presented above, a more nuanced analysis taking into account these specific factors would be essential for accurate predictions. In any case, this assessment does not seek to determine the direct impact on electricity prices. Instead, it focuses on analyzing the hypothetical scenario of such an increase and the extent to which different household groups could be affected, acknowledging the complexity of the Georgian energy landscape. Following this approach, a hypothetical price increase scenario of 5% was used.

Model outputs: Distributional impact on income shares of individual household income groups

Figure 6 shows the extent to which the share of income currently spent on electricity by different household income groups could be affected based on the chosen price increase scenario. As expected, the lowest-income groups would be most affected by this potential increase in electricity prices. Income groups earning less than GEL300/month could see their current share increase by between 0.3% and 0.6%.[10]

---

8   International Energy Agency. 2023. *Georgia Energy Profile.* https://www.iea.org/reports/georgia-energy-profile/overview.

9   E. Livny and I. Galdava. 2012. Everything You Wanted to Know about Your Electricity Bill. ISET Economist Blog. ISET Policy Institute. https://iset-pi.ge/en/blog/619-everything-you-wanted-to-know-about-your-electricity-bill.

10   This is a static analysis that aims to assess the impact on the share of income currently allocated to electricity expenditure if only the price of electricity were to change.

Figure 6: Change in Share of Income in the Short Term, by Income Group

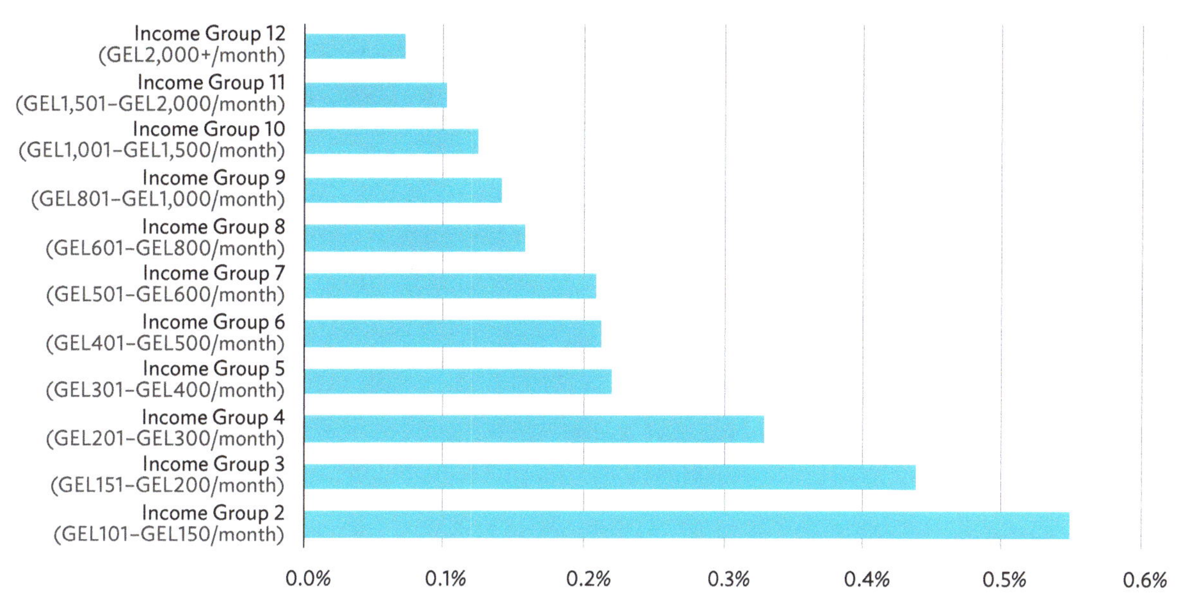

Source: Authors' own elaboration based on data from national household surveys, EnergoPro energy rates 2023, and estimates based on international literature.

In monetary terms, this represents less than GEL1/month for income groups earning less than GEL600/month and up to GEL2.2/month for the highest income groups. Figure 7 shows the potential monetary increase for each income group in the short term.

Figure 7: Potential Monetary Increase in the Short Term

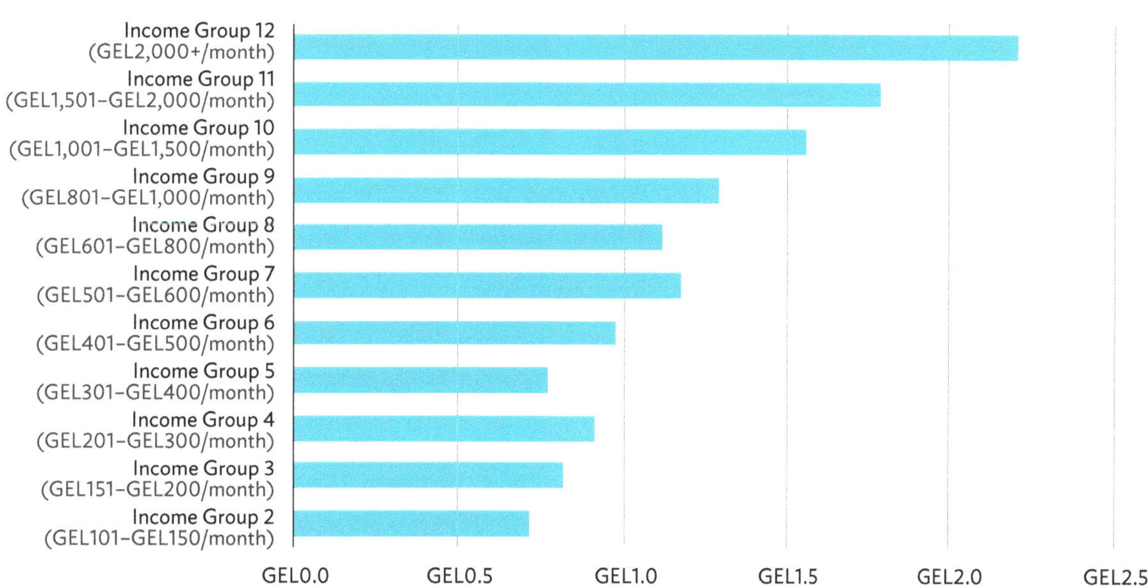

Source: Authors' own elaboration based on data from national household surveys, EnergoPro Energy rates 2023, and estimates based on international literature.

Figure 8 shows the distribution of the number of households in each income group.
As observed, the lowest-income groups who are potentially most affected by this price increase
(+0.2% in current share of income allocated to energy expenditure) represent a small portion
of the total population of Georgia (income groups earning less than GEL300/month add up
to 76,000, representing 7% of total households. This indicates that the measures planned to
mitigate the impact may be easier to manage and implement.

Figure 8: Distribution of the Number of Households in Each Income Group

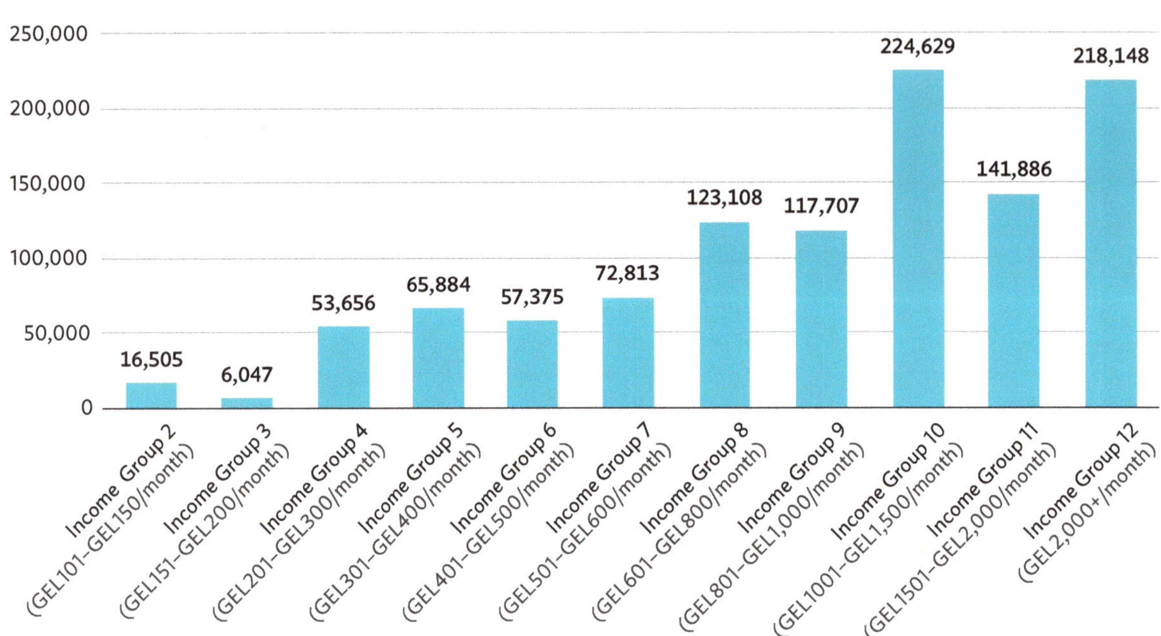

Source: Author's elaboration based on national survey.

Conclusions and Recommendations

**Due to the circumstances in Georgia described above, in particular the current low electricity
prices, the effects of a possible increase in energy prices are generally not as significant
for households. However, if electricity tariffs increase after the introduction of the open
market, an additional increase in the electricity tariff due to the introduction of renewable
energy sources may have more serious consequences.** In any case, it is crucial to recognize that
even if the impact is not substantial on average, there are some household groups that are already
economically vulnerable. For these vulnerable households, even a slight increase in expenditure
could have serious consequences. Government action should therefore be targeted at these
vulnerable groups. The segment comprising those most vulnerable and exposed to fluctuations in
energy prices encompasses a total of 76,000 households. This demographic constitutes 7% of the
total households in Georgia.

Gender and vulnerable groups

It is also important to recognize that women-led households are more vulnerable to an increase in energy prices due to the gender wage gap, which is estimated to be more than 30%.

With this in mind, the government should proactively plan and implement measures to address the potential challenges of this hypothetical price increase, which should target the group most vulnerable and exposed to energy price fluctuations. Possible measures could include the following:

(i) **Identify vulnerable households.** The first step is to identify who the vulnerable households are, where they are located, what their characteristics are, etc. This can be done through a top–down approach using national statistics in combination with bottom–up information from surveys and/or open inquiries with households.

(ii) **Dynamic subsidy programs.** Implementing dynamic subsidy programs that respond to market fluctuations and providing targeted financial relief to economically vulnerable households based on real-time energy price changes.

(iii) **Tax exemptions or reductions.** Taxes account for a large portion of the electricity tariff paid by consumers. To reduce the impact on vulnerable households without significantly affecting financial returns for market participants, the government can consider targeted fiscal measures for vulnerable households, such as value-added tax (VAT) exemptions.

(iv) **Flexible social safety nets.** Designing flexible social safety nets that can be adjusted according to market conditions ensures that vulnerable households receive timely and adequate support in response to fluctuating energy prices.

(v) **Integration into existing energy subsidy programs.** Integrate financial compensation for price increases into existing energy subsidy programs specifically designed for vulnerable groups.

(b)   Variables 2, 3, and 4: Job Creation for the New Renewable Energy Power Plants

Summary of Model Outputs

The construction and operation of renewable energy power plants, combined with strong policies and measures to promote local employment, can lead to a significant positive socioeconomic impact.

The number of jobs created was estimated using international averages per megawatt of installed capacity for each type of renewable energy and for each phase (construction and operation and maintenance [O&M]).

Jobs created during the construction phase of renewable energy power plants

Figure 9 shows the number of jobs created by the construction of the renewable energy power plants, in line with the construction scenario provided by the government. As observed, the construction of the renewable energy power plants will create almost 3,000 new job opportunities. This will result in an average of 780 people being employed in the economy each year from 2023 to 2029.

Figure 9: Number of Jobs Created for the Construction Phase of Renewable Energy Power Plants

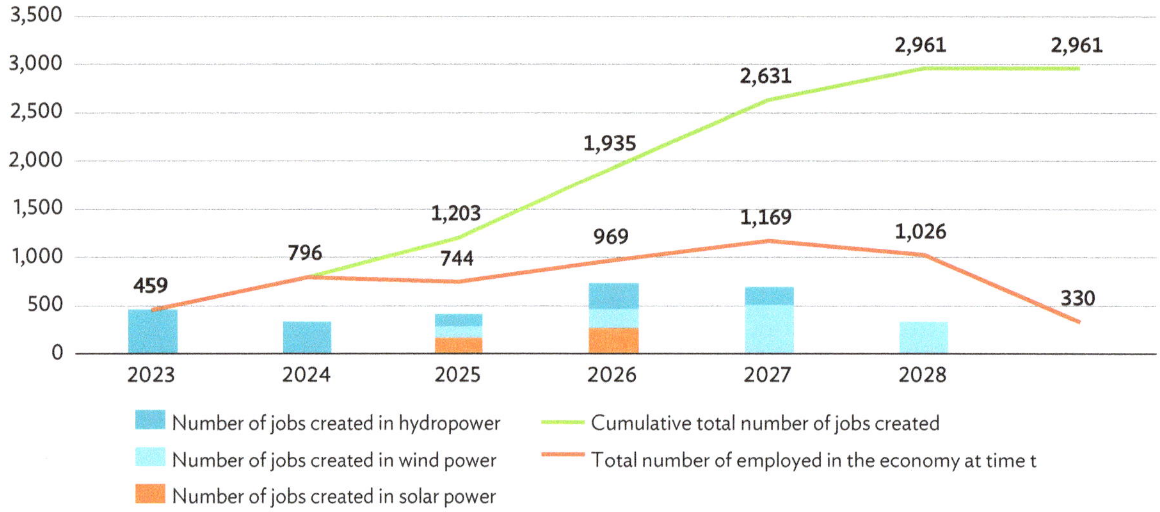

Source: Authors' own elaboration based on estimates from international literature.

It is worth noting that although these jobs are short term, i.e., lasting 1 to 2 years, they play a crucial role in providing workers with valuable experience. This experience not only contributes to their immediate economic well-being, but also positions them favorably for future employment opportunities in the ever-growing renewable energy sector as more power plants are built. Overall, the construction phase serves as a stepping stone that encourages the development of skills and expertise that can be used to further grow the renewable energy industry.

As presented in Figure 10, the total number of jobs created during the O&M phase (867) is significantly lower than that created during the construction phase (2,961). However, it is worth noting that these are permanent jobs that can be sustained for 20–25 years (the time frame of this assessment extends to 2035).

Figure 10: Number of Jobs Created for the Operation and Maintenance Phase of the Renewable Energy Power Plants

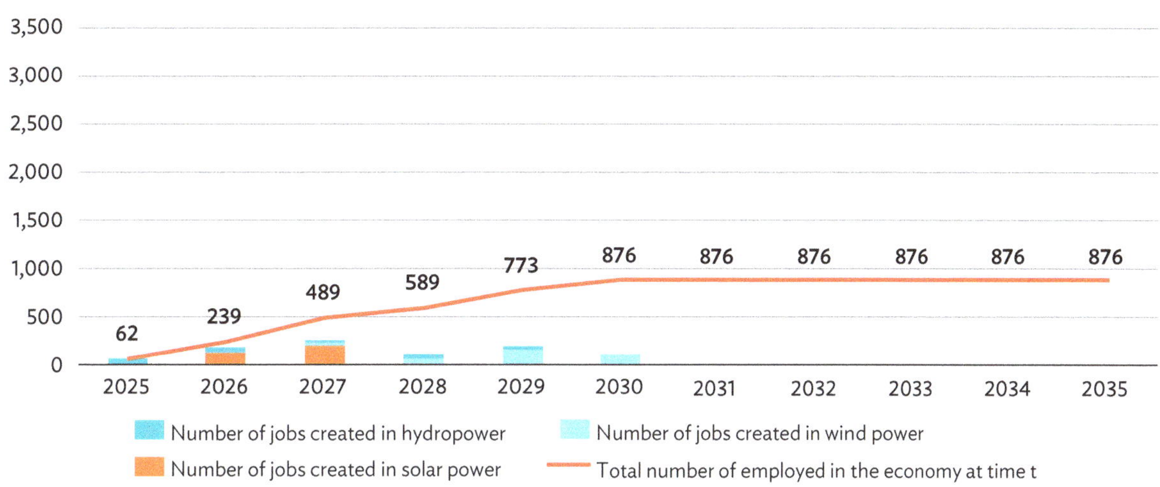

Source: Authors' own elaboration based on estimates from international literature.

Figure 11 shows the job opportunities created per region versus unemployment rate.

Figure 11: Job Opportunities Created per Region Versus Unemployment Rate

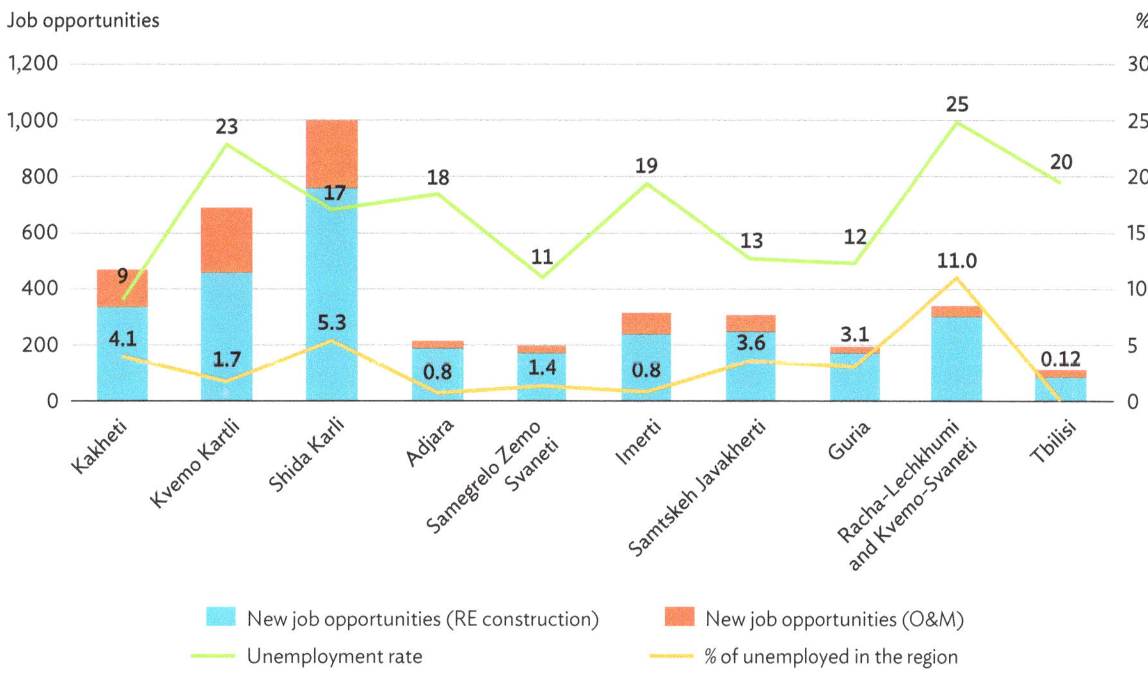

O&M = operation and maintenance, RE = renewable energy.
Source: Authors' own elaboration based on data from Geostat and outputs obtained from the assessment.

The model outputs obtained in the two variables mentioned above were contextualized based on the unemployment rate of each region where the construction of these plants is planned. The results show that in some regions with a high unemployment rate, such as Shida Kharli (17%) and Racha-Lechkhumi (25%), the number of jobs created in these regions accounts for a significant portion of the total number of unemployed (5.3% and 11%, respectively).

Conclusions and Recommendations

The assessment of job creation potential in the construction and operation phases of renewable energy power plants in Georgia shows an opportunity for economic growth and employment. Almost 3,000 jobs are expected to be created in the construction phase alone, which means an average of 780 job opportunities per year in the period from 2023 to 2029. While these jobs are temporary and last 1 to 2 years, they play a pivotal role as they not only promote immediate economic well-being, but also in equipping workers with valuable skills that qualify them for future employment in the emerging renewable energy sector.

With 867 permanent jobs, the O&M phase provides sustainable employment over the 20–25-year life cycle of the power plants. The disparity between the number of jobs in the construction and O&M phases underscores the temporary nature of jobs in the construction phase, but also emphasizes the potential for stable, long-term employment in the operational phase.

Gender and vulnerable groups

Vulnerable regions and communities. Contextualizing job creation based on regional unemployment rates underscores the potential of renewable energy projects in regions with high unemployment, particularly in Shida Kharli and Racha-Lechkhumi.

The jobs created not only alleviate local unemployment but can also contribute significantly to regional economic development. If a significant proportion of the unemployed population finds employment through these projects, the positive impact on these communities will become a key driver for sustainable growth.

Improving the employment rates can also have a positive impact on gender-specific issues such as domestic violence, which is associated with high unemployment rates.[11] This is a particularly important issue in Georgia where, according to UN Women, one in seven women between the ages of 15 and 64 say they have been victims of domestic violence.[12]

Gender disparities. The employment opportunities created by these projects can inadvertently perpetuate gender disparities. According to the Survey of the Current Demand and Future Intentions for Labour and Skills in the Transport and Energy Sectors by MOESD, the energy sector

---

[11]   CAGE Research Centre. 2021. Unemployment Substantial Increases Domestic Violence, New Study Finds. University of Warwick. https://warwick.ac.uk/fac/soc/economics/research/centres/cage/news/05-10-21-unemployment_substantially_increases_domestic_violence_new_study_finds/.

[12]   UN Women Georgia. 2020. Violence Against Women in Georgia. https://georgia.unwomen.org/en/digital-library/publications/2020/12/violence-against-women-in-georgia.

has a significant gender imbalance, with men largely dominating this sector. Women make up only 5%–7% of those studying energy-related subjects at higher education institutions and 17% of total employment in energy jobs.

To fully capitalize on the employment opportunities presented by renewable energy projects and address the associated challenges, the implementation of targeted measures and policies is crucial. Recommendations include the following:

(i)   **Regional skills audits, development programs, and engagement initiatives.** According to the MOESD's Survey of the Current Demand and Future Intentions for Labour and Skills in the Transport and Energy Sectors, the majority of the labor force in the energy sector is from Tbilisi (83%). To ensure that the most economically vulnerable regions benefit from these job opportunities, it is crucial to develop skills development programs at the regional level. It is also important to develop community engagement initiatives to inform residents of the potential job opportunities and facilitate their participation in the renewable energy workforce. This can include workshops, information events, and collaboration with local institutions.

(ii)   **Training and capacity building for women and diversity initiatives.** Introduce training and capacity-building programs specifically for women to address the skills gap identified in the survey. Providing tailored educational opportunities can increase the representation of women in the energy workforce. This should be accompanied by diversity and inclusion initiatives to encourage and empower women to pursue careers in renewable energy. This can include scholarships, mentorship programs, and awareness campaigns that challenge gender stereotypes.

(iii) **Local employment preferences.** Encourage the hiring of locals, especially in regions with high unemployment, during both the construction and O&M phases. Establish frameworks that prioritize local employment to maximize the positive impact on the regional economy.

(iv)   **Continuous skills development programs.** Implement skills development and training programs during the construction phase to improve the capabilities of the local workforce and ensure that they are well-equipped for future opportunities in the renewable energy sector when new renewable energy power plants are built.

(v)   **Transformational investment in education.** The government can collaborate with educational institutions to align their curricula with the skills requirements of the renewable energy industry. This will ensure that there is a continuous supply of skilled workers who can contribute to the sector.

By implementing these recommendations, the government can not only increase the positive impact of renewable energy projects on regions with high unemployment, but also create an inclusive and diverse workforce that maximizes the potential of all people, regardless of gender.

(c)   Variable 5: Impact on Property Value

Summary of Model Outputs

The construction of renewable energy power plants brings a spectrum of potential problems for local communities. These include aspects such as land use rights, forced migration, noise pollution, health and environmental problems, landscape aesthetics, and cultural disruption.

Quantifying these challenges in a socioeconomic model is intricate due to their subjective and case-specific nature and often requires an assessment at the project level. It is important to note that the results are only indicative and the impact on property values can vary from one location to another, with the specifics of each situation playing an important role. Property value assessments often involve a combination of objective factors (such as distance to the plant) and subjective factors (such as individual preferences and perceptions).

In this assessment, a global indicator is used—specifically, the change in housing and property value—which serves as a comprehensive measure. Property value, while not flawless, is chosen as it is assumed to cover a range of exogenous environmental, economic, and social factors that may have a direct or indirect impact on the property and thus comprehensively reflects the challenges posed by the construction of renewable energy power plants.

The average impact on property value is based on international studies and depends on the type of renewable energy being built.

(i)   Hydropower plants

Properties that could be affected by the impact include those located in the immediate vicinity of the hydropower plant and those located in close proximity to the downstream river. The impact on property value can be positive or negative:

- Negative impacts: Noise pollution, visual impact, safety concerns, land use, land access, downstream impacts (such as changes in water level, impact on economic and recreational activities, risk of flooding).

- Positive impacts: If the power plant is well designed and focuses on mitigating any negative impacts and maximizing the positive impacts (job creation, recreational areas, infrastructure development, flood risk reduction, etc.), the power plant can ultimately have a positive impact on property values.

In Georgia, hydropower plants—especially large power plants—are mostly perceived negatively by civil society, mainly due to the lack of adequate measures to mitigate the above-mentioned negative impacts. In such a case, the value of the affected properties could fall by 7% according to general estimates.[13]

(ii)   Wind power plants

The potential impact of wind power plants on property values is primarily determined by the visual impact and noise pollution. This mainly affects properties at a visible distance from the wind farm (2-kilometer radius) and can reduce their value by 12%.[14] In some cases, property values can be positively impacted if noise pollution is mitigated and properties benefit from lower energy costs.

Although there have been no significant challenges associated with wind power plants in Georgia to date, the success of the plan to build wind power plants may depend on effectively addressing the adverse effects and maximizing the positive ones.

(iii)   Solar power plants

Solar power plant farms seem to have the least significant impact on property values, although they are not negligible, which is why it is important to consider them.

In Georgia, the insufficient presence of utility-scale solar farms makes it difficult to precisely estimate the potential local issues. However, international studies estimate that these farms could affect property values by up to 1.5% within a 1-km radius.[15]

Model outputs

Figure 12 illustrates the potential impact on property values resulting from each type of renewable energy project within the affected properties. Naturally, the absolute value of the potential impact is higher in regions where the average price per square meter is more expensive, such as Tbilisi. This gives an indication of the compensation that can be given to local households, especially in the event that negative impacts are not adequately addressed by the project developers.

---

[13]   Author's own estimation based on various sources.

[14]   London School of Economics and Political Science (LSE). 2014. Gone with the Wind: House Prices Are Negatively Affected Where Turbines Are Visible. Blog. https://blogs.lse.ac.uk/politicsandpolicy/gone-with-the-wind/.

[15]   D. Gearino. 2023. Do Solar Farms Lower Property Values? A New Study Has Some Answers. https://insideclimatenews.org/news/15032023/solar-property-values/#:~:text=A%20new%20study%20finds%20that,just%20a%20little%20farther%20away.

Figure 12: Potential Impact on Property Value for Properties in the Immediate Vicinity, by Region
($)

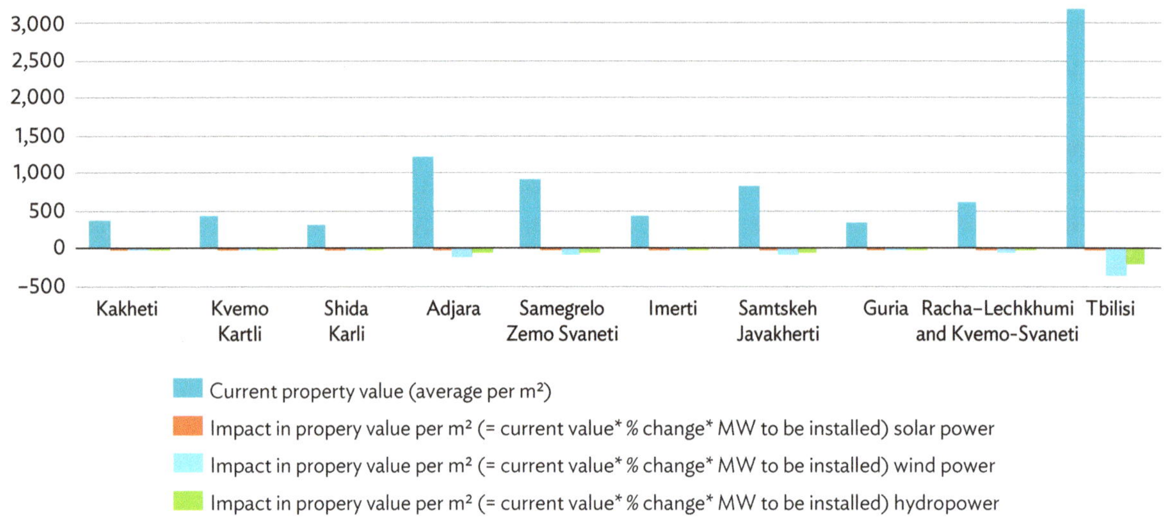

m² = square meter, MW = megawatt.
Source: Authors' elaboration based on SS.ge data and estimates based on international literature.

Conclusions and Recommendations

In conclusion, the construction of renewable energy power plants has potential for negative impacts on social, environmental, and economic aspects, which in turn affects property values, especially in the case of wind and hydropower projects. The case of hydropower plants in Georgia is a clear example of how the mismanagement of adverse effects has fueled a negative opinion of civil society, which has created impediments to both project success and the overarching national plan to promote renewable energy plants.

A key factor for success lies in the effective management of these impacts. If project developers take appropriate measures and these are supported by government policies, regulations, and measures, the construction of renewable energy power plants can have a positive impact on local communities and properties. The following proposed recommendations are intended to provide government authorities with tools to address this issue:

(i)   **Comprehensive social and environmental impact assessments** at project level are essential tools that provide insights into the nuanced challenges and guide targeted mitigation strategies.

(ii) **Offering financial compensation to local homeowners** as a preventive rather than a reactive measure would put Georgia at the forefront in this regard. Indeed, to date, this type of compensation is usually granted as a reactive measure, for example after a lawsuit. This compensation can be offered in a blended form with the support of project developers, the government, and agencies such as the multilateral development banks (MDBs). This can help to mitigate potential conflicts and build community trust.

(iii) **Allocating a percentage of tax revenues that remain in local communities** provides a direct and tangible benefit that promotes local development and mitigates potential negative impacts on property values. There is an ongoing regulatory process in Georgia that could support such a measure.

(iv) **Local benefits.** Facilitating a mutually beneficial relationship between a renewable energy project and the local community is essential to promoting positive engagement. This could be achieved through various measures, such as discounts on local energy rate, subsidies for property upgrades (to mitigate issues such as noise pollution) or even allowing residents in the vicinity to purchase ownership stakes in the renewable energy project. The latter not only provides economic incentives, but also encourages a sense of ownership and active community involvement.

(v) **Requiring project developers to invest in social projects,** such as recreational areas and infrastructure development not only offsets negative impacts, but also fosters a symbiotic relationship between energy initiatives and community well-being.

(vi) **An improvement in the regulatory framework** is necessary to comprehensively address negative environmental and social impacts. This also includes measures to tackle problems such as water and noise pollution.

(vii) **Concerns about land use and land access need to be addressed** by requiring project developers to carry out project-level assessments and plan specific mitigation measures. This was one of the main concerns of the CSOs interviewed during consultations. Addressing these concerns effectively is key to the successful implementation of such measures. During the stakeholder consultation, CSOs emphasized the need to define regulations with clear criteria that determine which types of land are suitable for solar, hydro, or wind projects. These criteria should be established in consideration of land use and access issues.

(viii) **Community engagement.** Encourage transparent and comprehensive communication with local communities and CSOs to address concerns, gather feedback, and involve residents in decision-making processes.

By integrating these recommendations into the planning and implementation of renewable energy projects, the government—in collaboration with other stakeholders such as project developers, CSOs, and MDBs—can contribute to a more sustainable and community-friendly renewable energy landscape.

(d)   Variables 6, 7, 8, and 9: Fiscal Revenues

Summary of Model Outputs

Although it may not seem immediately relevant from a just transition perspective, the assessment takes into account the positive impact on government revenue and recognizes its potential as a source of funding for the implementation of the recommended measures for this activity. In addition, a portion of this revenue could be channeled to regional and local authorities, which could make an important contribution to the socioeconomic development of the region.

The impacts considered are

    (i)   government subsidies,

    (ii)   government tax revenues from new RE generation,

    (iii)   government nontax revenues from new RE generation, and

    (iv)   government tax revenues from income tax on new workers.

Government subsidies

First, the qualitative assessment identified a potential impact on government subsidies. This could include both an increase in subsidies for renewable energy developers and a reduction in subsidies for other sources of renewable energy (see Chapter III. Qualitative Assessment).

However, following desk research and consultations with government counterparts, the assessment has concluded that the impact on this variable can be considered negligible. This is because the newly installed capacity is intended to meet rising energy demand and is not expected to replace current gas imports. Therefore, the activity will have no impact on existing government subsidies for imported gas, nor is it expected to have any other discernible impact on this aspect of energy activity. In addition, there are no plans to provide subsidies for renewable energy projects, as conveyed by government counterparts during the virtual meetings and confirmed during the in-country workshop.

Government tax revenues from new renewable energy generation

Figure 13 shows the potential additional tax revenue that the government could generate from the electricity produced by the new renewable energy power plants. This includes

- corporate income tax (15%), which is collected 100% by the government;
- VAT (18%), 100% of which is collected by the national state; and
- property tax (1%), 75% of which is collected by the municipalities and 25% by the national state.

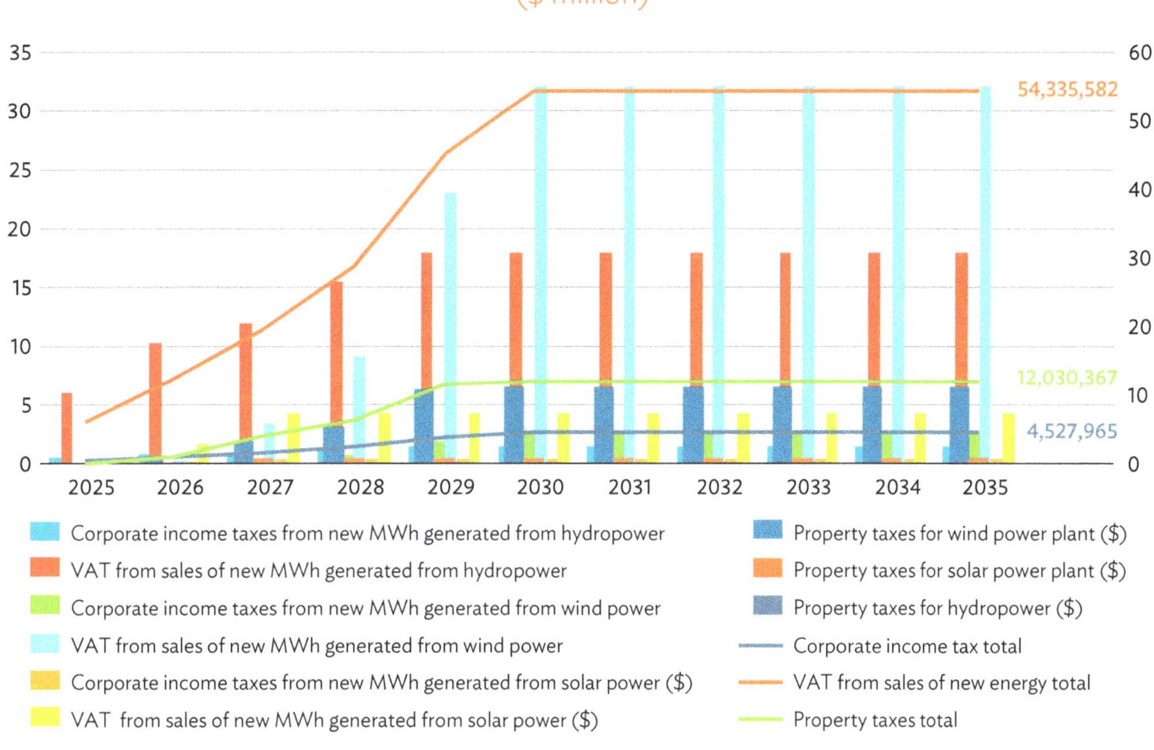

Figure 13: Tax Revenues from Renewable Energy Projects
($ million)

MWh = megawatt-hour, VAT = value-added tax.
Source: Authors' own elaboration based on data from Geostat and outputs obtained from the assessment.

As observed, VAT is the main source of tax revenue for the government at $54 million per year once all renewable energy projects are operational, followed by property tax ($12 million) and corporate income tax ($4 million).

It is worth noting that in Georgia, 75% of the total property tax is collected from municipalities, which represents an important source of revenue (up to $9 million per year once all power plants are operational) for local authorities, which can use it to finance measures to address specific regional issues.

Government tax revenues from new renewable energy power plants

Figure 14 shows the potential additional nontax revenue that the government could collect as a result of the new renewable energy plants. Solar power includes fees and permits, and lease payments.

As observed, once built, the plants would bring in a total of $1.3 million in additional revenue, mostly from fees and permits paid by hydro and wind energy power plants.

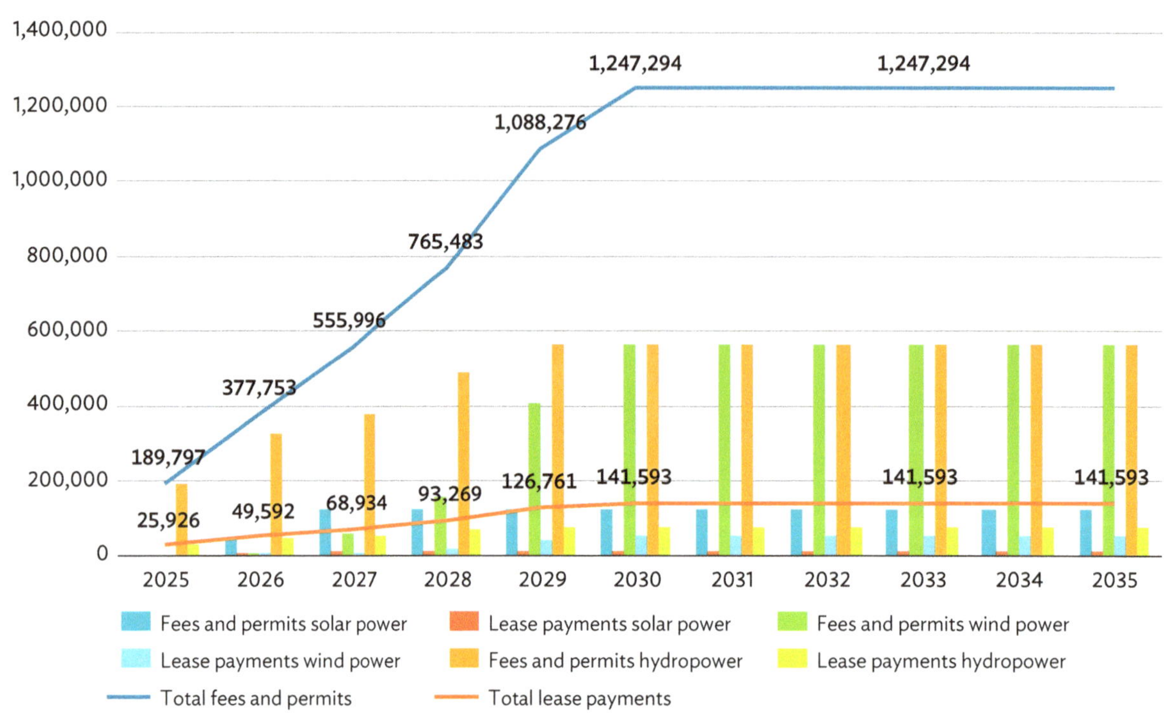

Figure 14: Nontax Revenue from New Renewable Energy Power Plants ($)

Source: Authors' elaboration based on national averages and estimates based on international literature.

Government tax revenues from new renewable energy power plant workers

Figure 15 shows the potential additional tax revenue that the government could collect from the new workers employed at the renewable energy power plants (both during the construction and the O&M phases). Income tax is calculated based on the average salaries of utility workers and a flat personal income tax of 20%.

Figure 16 shows the total potential fiscal revenue per year. Once all projects are operational, they have the potential to generate fiscal revenues of $74 million per year. Corporate taxes (corporate income tax, VAT, and property tax) account for by far the largest share (93% of total government revenue).

Figure 15: Income Tax Revenues from New Renewable Energy Power Plant Workers ($)

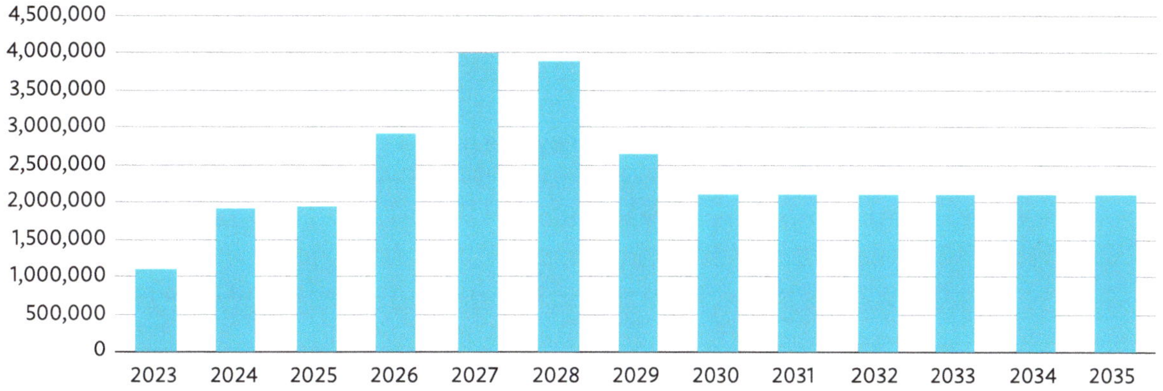

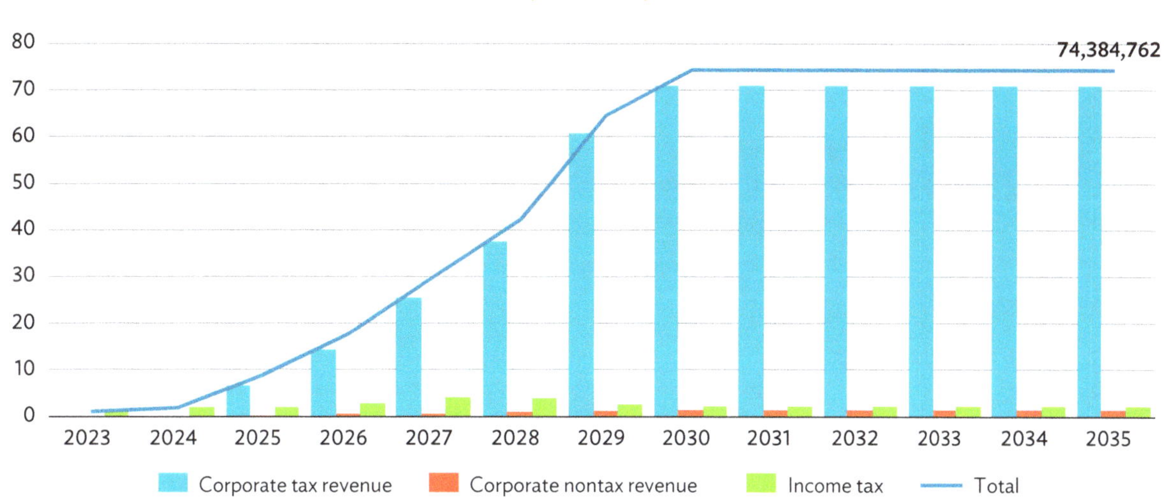

Source: Authors' own elaboration based on the average salary of a technician level employee in Georgia.

Figure 16: Total Government Revenue per Year ($ million)

Source: Authors' elaboration based on government revenue per year.

Conclusions and Recommendations

The evaluation of the potential fiscal revenues resulting from new renewable energy power plants reveals their role as a substantial revenue stream for the government that can be used to fund the measures described in the variables above. To achieve this, it is crucial to do the following:

(i)   **Establish a dedicated just transition fund.**

     Create a dedicated fund to collect government revenues from the new renewable energy power plants. This fund should be earmarked for predefined just transition initiatives related to mitigating negative impacts and maximizing positive impacts during the transition, prioritizing investment in local communities and transformational change.

There are several ways to feed this fund. A predetermined percentage of tax revenues from renewable energy projects can be set directly to the fund. Public–private partnerships can also be explored to leverage additional resources for just transition initiatives. Other stakeholders such as MDBs and private companies may also be interested in contributing to such a fund.

The design of such a fund can be guided by the model shown in Figure 17,[16] which serves as a tool to overcome the financing barriers faced by the Just Energy Transition Partnership, the world's largest financier of energy transitions.

Figure 17: Financing the Just Component of the Just Energy Transition Partnership Through a Just Transition Fund

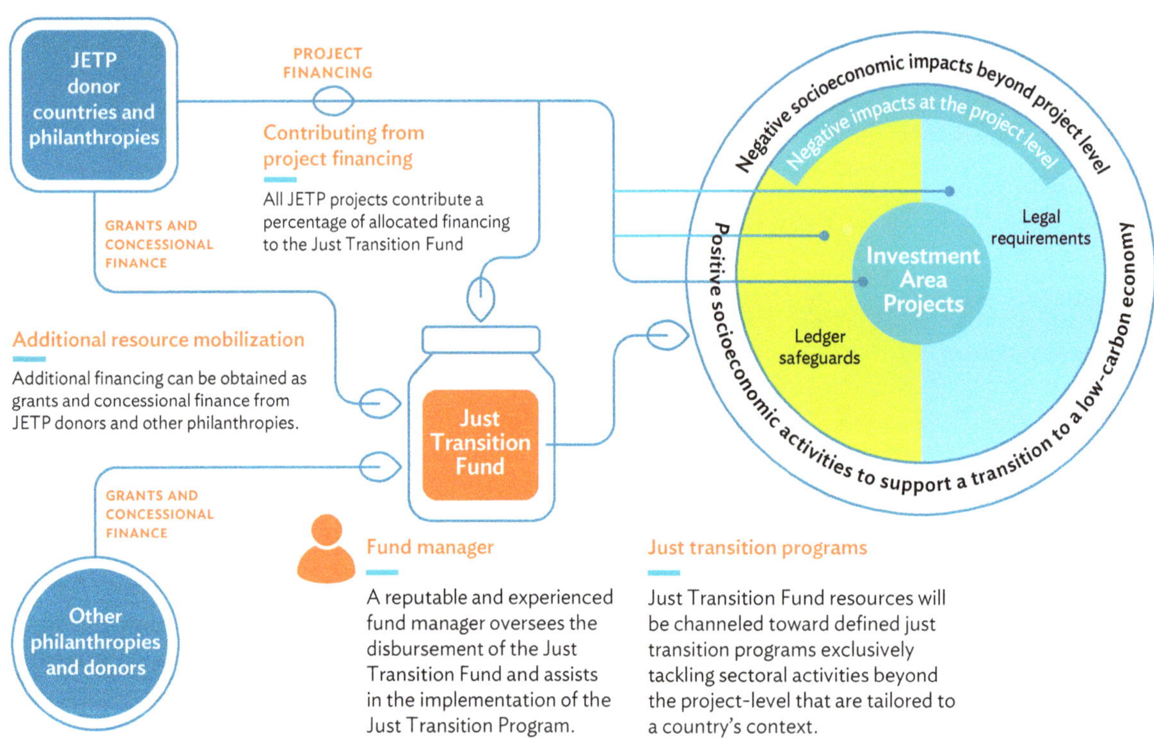

JETP = Just Energy Transition Partnership.
Source: Adapted from Neyen.

(ii)   **Promote the collection of fiscal revenues by local authorities.**

Georgia has already implemented certain mechanisms and regulations to facilitate the collection of fiscal revenues by local administrations. This applies in particular to property tax and an ongoing renewable energy regulation that is in the pipeline.

---

[16]   Neyen. 2023. Financing the Just Component of JETP Through a Just Transition Fund. https://neyen.io/financing-the-just-component-of-jetp-through-a-just-transition-fund/.

The promotion of such regulations and mechanisms is a crucial lever for the effective implementation of just transition measures. Given that local authorities have a nuanced understanding of local issues, circumstances, and challenges, this approach guarantees a more targeted and effective use of the revenues collected. By targeting fiscal resources to the unique needs of each community, particularly with regard to just transition initiatives, these regulations empower local governments to address challenges and promote positive outcomes in a way that resonates with their specific circumstances.

It is imperative that the government provides guidance and capacity-building support to ensure that allocated funds are utilized optimally. This may include providing a positive and negative list of measures that can be financed with this fiscal revenue. Transparency and robust reporting mechanisms also play a pivotal role in this process. They promote accountability and trust between the central and local levels of government and society as a whole.

(e)   Recommendations on Activity 1

Table 10 summarizes the key recommendations for each measure based on the impacts identified and the corresponding significance assessment.

The details of the significance assessment can be found in Appendix 2.

Table 10: Summary of Findings, Recommendations, and Impact Significance

Variable	Type	Significance	Impact Summary	Recommended Measure
Variable 1: Distributional impact on electricity expenditure in household income groups	RISK	MINOR	>0.2% change in the current share of income spent on energy for 7% of the total households (~76,000), representing the most economically vulnerable groups. Intensified impact in women-led households due to wage gap (+30%).	• Identify vulnerable households, including vulnerable women-led households. • Targeted financial support for the lowest and most vulnerable household groups (estimated budget GEL76,000/month) to be included in existing support measures.
Variables 2–4: Job creation for the new RE power plants	OPPORTUNITY	MODERATE	~4,000 job opportunities created in RE sector	• Provide incentives for companies in the RE sector to create local jobs. • Develop regional training and skilling programs and specific programs for women.

continued on next page

Table 10 *continued*

Variable	Type	Significance	Impact Summary	Recommended Measure
Variable 5: Impact on property value and local communities	RISK	MAJOR	Estimated average impact on property values in the immediate vicinity of RE power plants: –7% hydro, –12% wind, –1.5% solar and issues related to land access, land use, environmental and social issues, generating severe social acceptance issues.	• Improve the regulatory framework to address negative environmental and social impacts. • Mandate project developers to invest in local social projects. • Tackle land use and land access concerns through regulation. • Allocate a percentage of tax revenues to remain in local communities for investment in local social projects.
Variables 6–9: Fiscal revenues	OPPORTUNITY	MODERATE	~$72 million per year (mainly corporate tax revenues)	• Use these revenues to improve opportunities and mitigate negative impacts through a just transition fund and/or just transition guidelines. • Promote the collection of taxes by regional administrations and stream the use of these revenues toward just transition measures.

RE = renewable energy.
Source: Author's elaboration.

The assessment found that the biggest challenge in implementing renewable energy projects is the impact on local communities (reflected in the impact on property values). This is mainly due to the negative public perception of renewable energy in the country, which is particularly intensified by past experiences in the hydropower plant sector. If this skepticism is not properly addressed, it could pose a serious challenge to the successful implementation of planned renewable energy projects. Therefore, it is crucial that the socioeconomic implications identified in this study are carefully considered to mitigate the negative impacts and maximize the opportunities.

On the other hand, the assessment has shown that the potential increase in electricity prices as a result of the introduction of these new renewable energy sources is unlikely to significantly affect the purchasing power of most household income groups. Nevertheless, it is important to recognize that even a slight increase in the expenditure share could pose challenges for the most economically vulnerable groups (around 76,000 households). Therefore, a proactive plan with mitigation measures to address these potential economic repercussions should be implemented. Given the amount needed to mitigate the impact (about GEL76,000/month), the most effective measure would be to include this financial support in the existing financial aid mechanisms for these target groups.

The assessment has also shown that the establishment of new renewable energy power plants has the potential to open up opportunities. In particular, this includes the creation of a total of around 4,000 new jobs (during the construction and operation and maintenance phases).

This is a particularly important opportunity as some of these plants are to be built in regions struggling with high unemployment. In addition, the implementation of RE projects can generate an important additional source of revenue for public administration, which can be used to finance just transition measures.

In summary, the main measures recommended to mitigate the negative impacts and maximize the opportunities are as follows:

(i)   **Refine the regulatory framework to mitigate the negative impacts** of renewable energy projects. The main concerns raised during the public consultations include

  - environmental issues, especially for hydropower plants;
  - land issues: land use (e.g., for cattle grazing) and land access rights (e.g., if a power plant obstructs access to an adjacent forest); and
  - economic issues (property value, impact on economic activities downstream of the power plant).

(ii)   **Refine the regulatory framework to maximize positive impacts,** with a focus on promoting

  - incentives to RE companies to employ local workers;
  - the investment in social infrastructure;
  - skills development programs and special programs for the most vulnerable groups and women;
  - the sharing of benefits with local communities (preferential price for energy, ownership, etc.);
  - tax collection by local authorities;
  - the establishment of a just transition fund and/or a set of guidelines and recommendations on how fiscal revenues should be used to fund just transition measures; and
  - environmental benefits (for example, flood control for communities living downstream from a hydropower plant.

(iii)   **Improve communication with CSOs to enhance the public perception of renewable energy in Georgia.** The government should prioritize improving its communication with CSOs. A specific example that underscores this need is the concern expressed by CSOs about the government's focus on building new renewable energy plants instead of improving the efficiency of existing ones. Despite the prevailing sentiment that the current plants and distribution systems are not efficient enough to supply local households directly, the government takes a different perspective. According to official statements, the efficiency of energy transmission has improved significantly and is now only 5% to 6% energy loss during transmission. Furthermore, a study by the MOESD found that investments in the energy efficiency of existing power plants would result in a modest improvement of 5%. Given these contrasting views, it is clear that there is a communication gap between the government and CSOs.

This strategic approach not only addresses negative issues, but also cultivates an inclusive, sustainable, and supportive framework for the advancement of renewable energy initiatives. By diligently implementing these key actions, the Government of Georgia can actively promote a just transition where all stakeholders benefit from this transformation of the sector.

Activity 2: Energy Efficiency Reform for the Construction of New Buildings

1.   Description of Activity

The Georgian Parliament has passed a law on energy efficiency and energy performance in buildings that brings the country closer to European Union (EU) standards. The legislation aims to reduce emissions and pollution, improve the energy efficiency of buildings, reduce energy imports, and bolster the country's energy security. The reform stipulates that all construction permits issued in Georgia from July 2023 must meet a number of minimum energy efficiency requirements set out in the Law on Energy Efficiency. These include the requirements for the energy performance of buildings, parts of buildings or building elements, as well as the energy performance requirements for the engineering and technical support systems of a building.

One of the main goals stated in the 2030 National Climate Strategy and 2021–2023 Action Plan (CSAP) is therefore to support the development of low-carbon approaches in the building sector by promoting climate-smart and energy-efficient technologies and services.

This reform stipulates that all construction permits issued in Georgia from July 2023 must meet a set of minimum energy efficiency requirements defined by the national authorities. Project applications must include declarations confirming compliance with the new requirements. The Ministry of Economy provides training for architects and project designers on the declarations (which must be submitted to confirm compliance with the new requirements).

2.   Value Chain Map

Figure 18 shows the general value chain for energy efficiency reform in new buildings.

Figure 18: Value Chain Map for Energy Efficiency Activity in Georgia

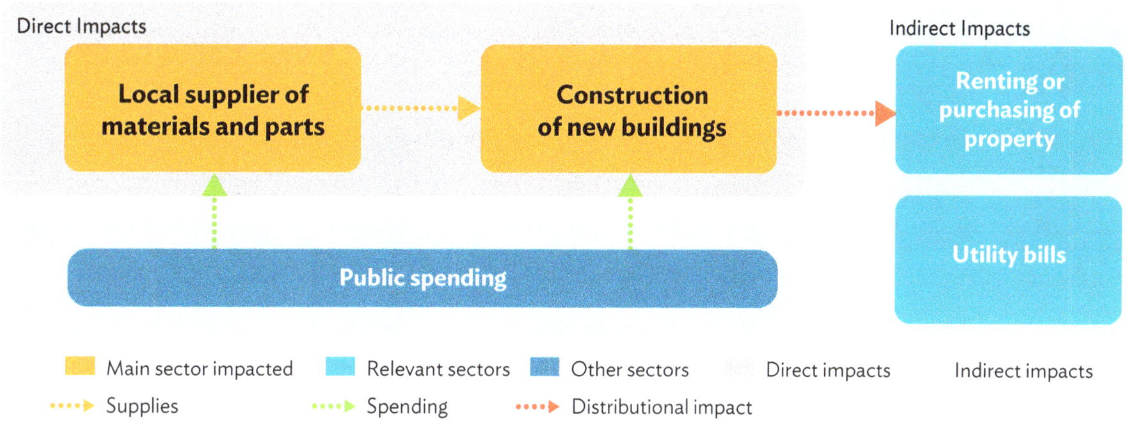

Source: Author's own elaboration.

3.   Qualitative Assessment: Identification of Affected Stakeholders and Risks and Opportunities

The primary impact of this activity seems to be the distributional effect on renters and homebuyers who will most likely face higher prices due to the new building standards. An expected increase in construction costs due to the reform will most likely lead to higher prices, which could compromise the ability of future renters and homebuyers to access these new buildings. Those who cannot afford the new buildings will therefore be forced to live in less energy-efficient buildings where they will have to pay more for utilities.

Another important issue is the capacity of micro, small, and medium-sized construction companies and suppliers to adapt to the reform at the same pace as other larger entities. These entities will face challenges related to regulatory and administrative burdens, as well as an increase in training, reskilling, purchasing, and operational costs.

The implementation of this measure is not expected to create a significant number of jobs. Although the way in which new buildings are constructed will be affected, no new jobs are expected as the buildings to which the energy efficiency reform applies would most likely have been built anyway, albeit to less energy-efficient standards. Those who will be working in the construction industry without the reform are those who will be working in the construction industry with the reform. Therefore, no additional job creation is expected. Suppliers of more energy-efficient materials are also expected to remain the same, although under stricter energy-efficiency standards.

The implementation of the reform will entail the creation of new jobs, but these are not considered significant enough to be included in this assignment.

4.   Variables Included in the Assessment

Following the qualitative assessment presented above and discussions with government counterparts, the socioeconomic variables listed in Table 11 were selected for the study.

Table 11: Types of Impact for Each Socioeconomic Variable

Variable	Type of Impact	Description
1 Distributional impacts[a] on constructor suppliers of construction companies	Indirect	Assesses the extent to which different size groups of constructor suppliers are affected by the reform. The focus is on micro, small, and medium-sized suppliers to construction companies that may face financial challenges in meeting the investment requirements associated with the reform.
2 Distributional impacts on building construction companies	Direct	Assesses the extent to which different size groups of construction companies could face financial challenges to meet the investment requirements associated with the reform.
3 Distributional impacts on renting costs	Indirect	Considers the impact of fluctuations in rental prices on the proportion of household income allocated by each income group to these expenses. Gender perspective: Vulnerable groups of women could be more affected due to the gender wage gap.
4 Distributional impacts on house acquisition prices	Indirect	Indicates the extent to which variations in house prices affect the proportion of household income that each income group allocates to those expenses. Gender perspective: Vulnerable groups of women could be more affected due to the gender wage gap.
5 Distributional opportunities thanks to energy savings	Indirect	Reflects the distributional opportunities that energy efficiency reform offers in terms of energy savings for each household income group as a percentage of their allocation. It also assesses the gap between the change in rental costs and energy savings.

[a]   While the term "distributional impact" is commonly associated with the effects on different population groups, in this assessment we extend its application to the impact on various groups with common characteristics (e.g., in this case, different size groups of suppliers). Source: Author's own elaboration.

Exclusions of the Assessment

No significant job creation is expected from the implementation of this activity. Although the way in which new buildings are constructed will be affected, no new jobs are expected as the buildings to which the energy efficiency reform applies will most likely still be built, albeit to less energy-efficient standards. Those who would have been working in the construction industry without the reform are those who will be working in the construction industry with the reform. Therefore, no job creation is expected. Similarly, suppliers of more energy-efficient materials are also expected to remain the same, although to stricter energy-efficiency standards.

5. Summary of Model Outputs and Recommendations

This subsection presents the key findings for each socioeconomic variable, including a summary of the model outputs and recommendations for government counterparts.

(a) Variable 1: Distributional Impact on Constructors' Suppliers

Summary of Model Outputs

The upcoming reform will significantly increase the demand for energy-efficient building materials. As a result, building suppliers will have to adapt by manufacturing these new materials. This will result in additional expenses, such as material costs, machinery costs, regulatory costs, and training costs.

Suppliers that are not able to meet these new requirements risk losing market share and ultimately possibly going out of business. This variable is used to identify local suppliers at risk, investigate the primary reasons for their vulnerability, and propose preventive measures.

The different sizes and revenues of suppliers were taken into account in the assessment. It is worth noting that larger suppliers may have greater resources to bear the associated costs of materials, regulations, and training, while smaller suppliers may experience a greater proportional impact on their operations. Including these distributional effects in the analysis provides a more comprehensive understanding of the disparities and facilitates a more nuanced assessment of the situation.

Number of building suppliers by size

Figure 19 illustrates the number of building construction suppliers in Georgia by size. As observed, the vast majority of geosuppliers consist of small and micro enterprises. In fact, most of the construction materials in Georgia are imported, especially the most processed ones. Local micro, small and medium-sized enterprises (MSMEs) produce essential "raw" building materials such as bricks, concrete blocks, steel reinforcement, and cement.[17]

---

[17] Primary data provided by MOESD during stakeholder consultations.

Figure 19: Number of Suppliers of Building Materials by Size

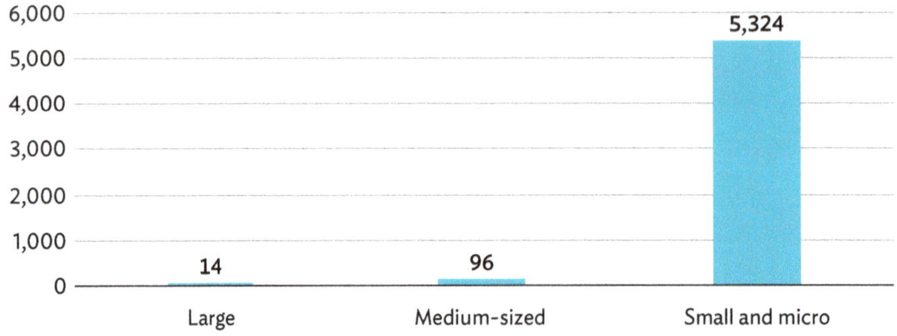

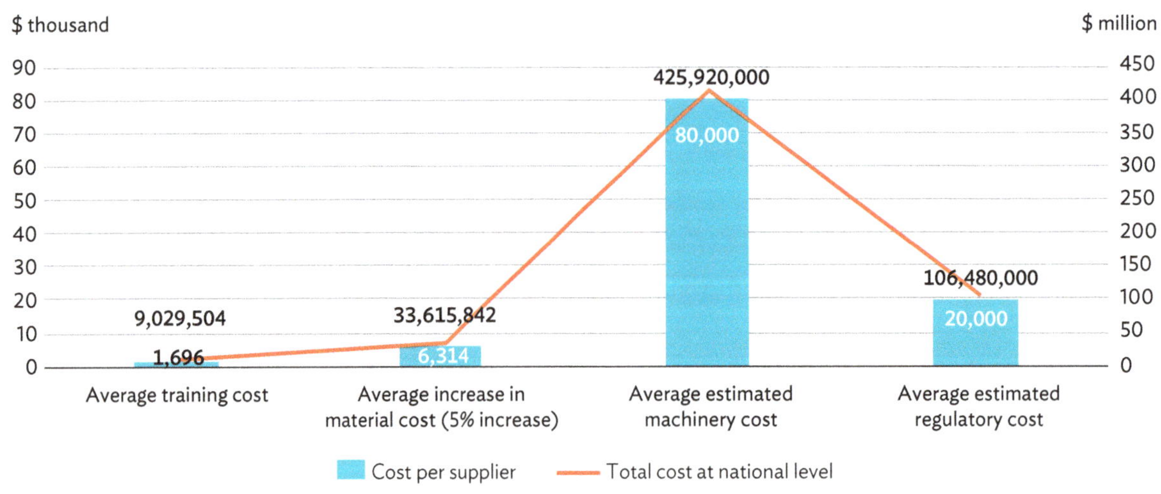

Source: National Statistics Office of Georgia. Geostat Database. https://www.geostat.ge/en (accessed 5 March 2024).

Estimated costs for micro, small and medium-sized enterprises

The assessment's approach, validated through stakeholder consultations with government counterparts and local CSOs, is based on the assumption that only small and micro suppliers face the risk of noncompliance with the reform. This assumption is based on the analysis of the financial reports of medium-sized and large suppliers,[18] which show that they are able to make the necessary investments to comply with the reform. Therefore, the cost breakdown presented in Figure 20 is specific to MSMEs.

Figure 20: Estimated Costs for Small and Micro Suppliers

Source: Authors' elaboration with estimates based on national and international literature, considering a business of 10 employees.

As can be seen in Figure 20, the biggest impact comes from the need to invest in machinery for the production of new, efficient materials. This is particularly noteworthy because of their upfront nature and the uncertainty as to whether these investments will be sufficient to meet the new demand.

---

[18]   Service for Accounting, Reporting and Auditing Supervision. Annual Statements Register. https://reportal.ge/en/Reports.

Even if these suppliers mainly produce basic construction materials, they would still need to invest in machinery to produce these new efficient raw materials. This could include, for example, aerated concrete block production lines, insulation manufacturing machines, green concrete production machines, etc. MSMEs should expect an upfront investment of $80,000 on average.[19] (Note: This is a general estimate and actual costs may vary greatly depending on factors such as the machine needed, the production capacity, level of automation, technological features, etc.).

Regulatory costs are another focal point. Certain elements such as fees for the certification process (including consulting/legal advice) and expenses related to permits and licenses are fixed costs that require an initial investment.

The cost of training, while also an upfront, fixed expense, does not appear to be a significant economic challenge for businesses.

Likewise, the increase in material costs is manageable as you can see in Figure 20. Although the increase may seem noteworthy, being a variable cost allows for direct recovery when the material is sold, which mitigates its impact and rendering it non-substantial.

The orange trend line delineates the total costs at national level for all 5,324 suppliers. It is crucial to note that the costs shown are indicative only and may vary depending on size, revenue, existing costs, regulatory compliance infrastructure, machinery capabilities, and other factors.

Conclusion and Recommendations

The assessment highlights the challenge faced by MSMEs in Georgia's construction industry as they grapple with energy efficiency reform. The upfront investment required for machinery to produce new, efficient materials is a significant hurdle for MSMEs in particular, raising concerns about their ability to meet the reform's demand requirements.

However, this also offers the opportunity not only to mitigate the negative effects, but also to stimulate and revitalize an underdeveloped sector in Georgia. If the appropriate policies and measures are put in place, this could boost the domestic production of energy-efficient construction materials, which today is largely dependent on imports. Some of these measures include the following:

(i) **Facilitate customized financing solutions.** This can be done through various financing instruments that can be implemented by the government and/or in collaboration with MDBs and the private sector. Such financing instruments include the following:

- **Preferential finance programs** with favorable terms to reduce the financial burden on MSMEs, especially for the purchase of new machinery, while promoting local economic development. Innovative ways to promote these programs can be explored to avoid significant financial repercussions. Some of these may include the following:

---

[19] Estimation based on an average of retail prices of the machines listed above.

- ☐ **By reducing energy consumption, the government can reduce spending on subsidizing gas.** The saved funds can be redirected to a financing arrangement with commercial banks that provide MSMEs with low-interest rates.
- ☐ **Since energy-efficient materials may be more expensive, the government can benefit from higher VAT revenues.** Part of this additional revenue can be used to support the above-mentioned preferential financing line to promote sustainable practices and economic growth.

- **De-risking mechanisms** to encourage financial institutions to support MSMEs. **Strategic investment in tailored capacity building and training support** is another essential initiative. These targeted efforts should aim to provide suppliers with a comprehensive understanding of the impact of the reform on their businesses, anticipate shifts in demand, and identify the necessary measures to sustain their operations. The first step should be to identify suppliers and delineate their characteristics, including factors such as age, gender, location, and vulnerability. This profiling would make it possible to provide specific and tailored capacity-building and training programs, ensuring a more effective and impactful support mechanism.

(ii) **Stimulate local demand for energy-efficient construction materials** to develop this sector and create economic growth and new employment opportunities. This can be achieved through

- **Financial incentives** for constructors that use a defined threshold of local suppliers (e.g., fiscal benefits);
- **Government procurement policies** for public projects;
- **Measures to improve the competitiveness of local MSMEs.** This can be done through tariffs on imported materials or financial benefits for local MSMEs; and
- **Capacity-building programs** could also aim to attract new talent to the industry, with a particular focus on underrepresented groups, including women.

(b)   Variable 2: Distributional Impact on Constructors

Summary of Model Outputs

After conducting desk research and several consultations with government counterparts and CSOs, it appears unlikely that the energy efficiency reform will have a significant negative impact on constructors, including MSMEs. Several key factors contribute to this nuanced assessment:

Impact on costs

Cost pass-through: The anticipated increase in material and regulatory costs is expected to be met through an adjustment in the selling price of the building (estimated at 5%).[20] This mechanism safeguards the financial viability of constructors amid rising operational expenses.

---

[20]   Primary data provided by MOESD.

Impact on demand

A price increase may raise concerns about a potential drop in demand that could affect constructors, especially MSMEs that may not have enough funds to transition over a period of low activity. However, the local context in Georgia and recent data contradict this apprehension. While house prices recorded a notable increase in 2021 (+7%, according to CEIC data), demand for houses also increased, partly due to increased interest from foreign investors and limited supply.[21] This suggests that the market remains resilient overall, although there are fluctuations in demand within certain income groups.

A study conducted by the University of Cambridge found that energy efficiency levels have a positive effect on house prices, with an average increase of 6% for each Energy Performance Certificate band improvement. However, the study found no evidence of a negative impact of energy efficiency on housing demand, measured by the number of transactions or time on the market. **In addition, some studies from other countries have concluded that there is no evidence of a negative impact of energy efficiency on housing demand,** even if these measures imply a price increase:[22]

Diverse market impact and distributional impact on accessibility

While it is acknowledged that not all income groups will be able to maintain the same level of access to new properties, the impact of the diversified market ensures that constructors can adapt to varying demand patterns. The presence of different buyer segments, including foreign investors, provides a layer of stability.

However, this means that certain groups of the population may face a new hurdle to purchase a new house. This is further studied in the next variables.

Conclusion and Recommendations

In summary, the evidence suggests that the reform's impact on constructors, including MSMEs, is likely to be mitigated by their ability to adjust selling prices in response to the increased costs and by high and diversified demand. This contributes to a scenario where constructors can effectively navigate potential challenges. However, it is important to monitor these trends closely and remain responsive to changing market conditions.

[21] L. C. Delmendo. 2024. Georgia's House Price Growth Continues to Accelerate. Global Property Guide. https://www.globalpropertyguide.com/asia/georgia/price-history.

[22] F. Fuerst and H. Adan. 2017. Do House Prices and Rents in the Private Rented Sector Reflect Energy Efficiency Levels? BEIS Research Paper. No. 2020/027. Department for Business, Energy & Industrial Strategy, University of Cambridge. https://assets.publishing.service.gov.uk/media/5f3e8c7ce90e072ed1ab1b00/beis-cambridge-house-price-report.pdf.

(c)   Variable 3: Distributional Impact on House Acquisition Affordability

Summary of Model Outputs

As mentioned earlier, the increase in construction costs is expected to translate into a subsequent uptick in house prices. According to MOESD, this increment is expected to be about 5% compared to current market prices.

The overall elasticity factor for house acquisition in Georgia appears to be relatively low, meaning that the quantity demanded of housing is not very sensitive to price changes. Indeed, despite price increases in recent years, the housing market remains robust, buoyed by strong demand and limited supply (footnote 21).

However, this does not mean that all household groups show this low elasticity to price increases. While for high-income households this has no impact on their accessibility to acquire a property, this is not the case for low-income households, for whom such price increments could be a significant barrier to housing accessibility.

Assessing the potential implications of a 5% surge in prices on the accessibility of various household income segments is a significant issue from a just transition perspective. The importance lies not only in the accessibility of housing, but also in the potential to widen disparities between high-income and low-income groups: high-income households would likely secure access to energy-efficient properties, while low-income households may be limited to less efficient alternatives, assuming an additional burden of increased energy expenses. In the context of a just transition, it is important that climate measures such as energy-efficient housing benefit all segments of the population. Currently, energy prices in Georgia are not a major concern for people as the government regulates and controls prices. However, in the foreseeable future, the transition to an open market structure could precipitate elevated prices, potentially giving rise to a noteworthy issue. Therefore, it is important to carefully consider this potential impact.

In this framework, this variable assesses the extent to which different household income groups might be affected by this price increase. It also explores different ways in which the government can support lower-income households and facilitate their access to such housing, thus promoting inclusivity and addressing potential disparities.

Distributional impacts on the affordability of house acquisition

Figure 21 illustrates the current allocation of income for mortgages by different household income groups compared to the maximum share permitted for mortgage loans according to national regulations. In 2019, Georgia introduced a regulatory framework that prohibits banks from issuing mortgage loans that exceed a certain percentage of the borrower's income, with a maximum loan term of 15 years. For the purposes of this assessment, these regulatory thresholds are considered indicative of a prudent share of expenditure in mortgage loans, designed to safeguard essential expenditures such as that on health and education.

Figure 21: Maximum Allowable Share of Mortgage Loans in Income, by Income Group

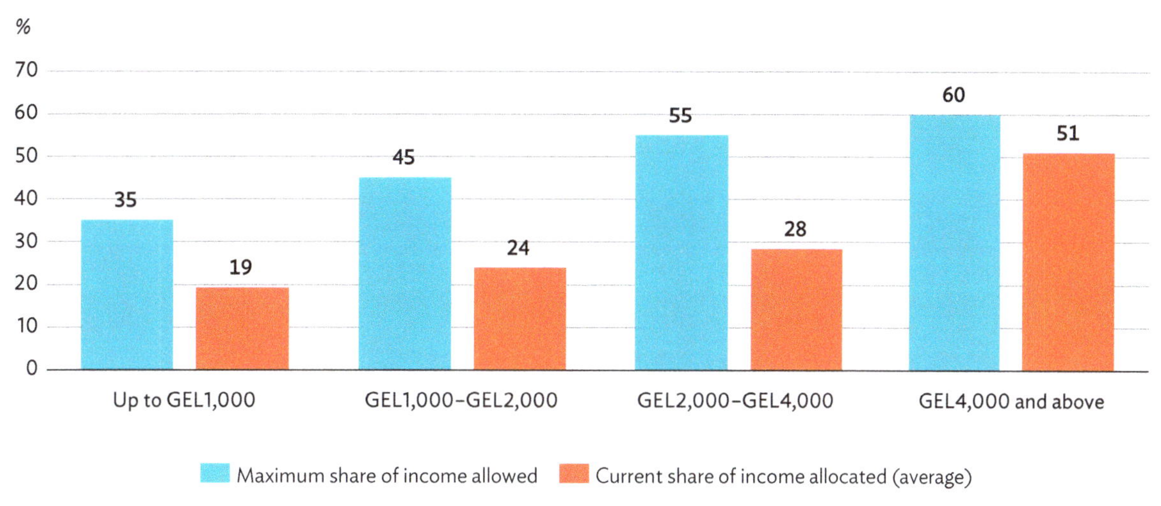

Source: Authors' elaboration based on data from Global Property Guide and Household Surveys.

As can be observed, lower-income groups spend a significantly smaller proportion of their income on mortgage loans. This is primarily due to the fact that for households with limited financial means, this expenditure is primarily used to secure basic housing needs, to which the majority of the Georgian population has access, as Figure 24 shows). In contrast, higher-income groups might pursue more comfortable living arrangements or view real estate as an investment commodity. Their housing goals go beyond mere necessity and reflect a broader spectrum of financial aspirations and consequently a higher share of income. Nonetheless, in each of these categories, a 5% increase in housing prices suggests that their ability to meet this prudent share of expenditure may be at risk.

Gender perspective

It is important to consider the potential impact of this price increase on women looking to purchase a house. Although a comprehensive analysis is limited due to the limited data available, the existing gender wage gap in Georgia, which currently stands at +30%, presents a substantial and notable challenge for women. This significant pay disparity underscores the additional hurdles that women can face in the housing market and highlights the importance of addressing gender-based economic disparities to ensure equal access to housing opportunities. Figure 22 shows the disparity between women and men, taking into account the gender income gap (30%) in a scenario where expenditure on housing acquisition would be equal.

Figure 22: Gap Between the Share of Income Allocated by Men and Women to Mortgages

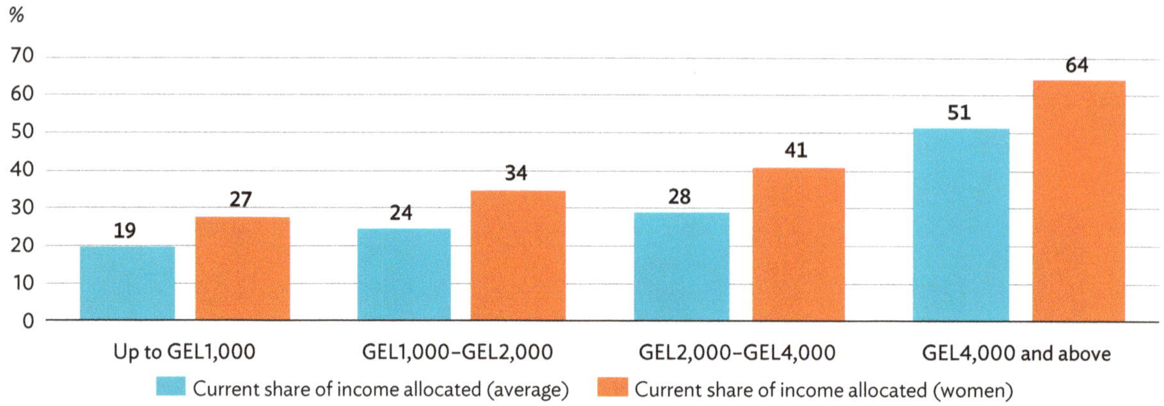

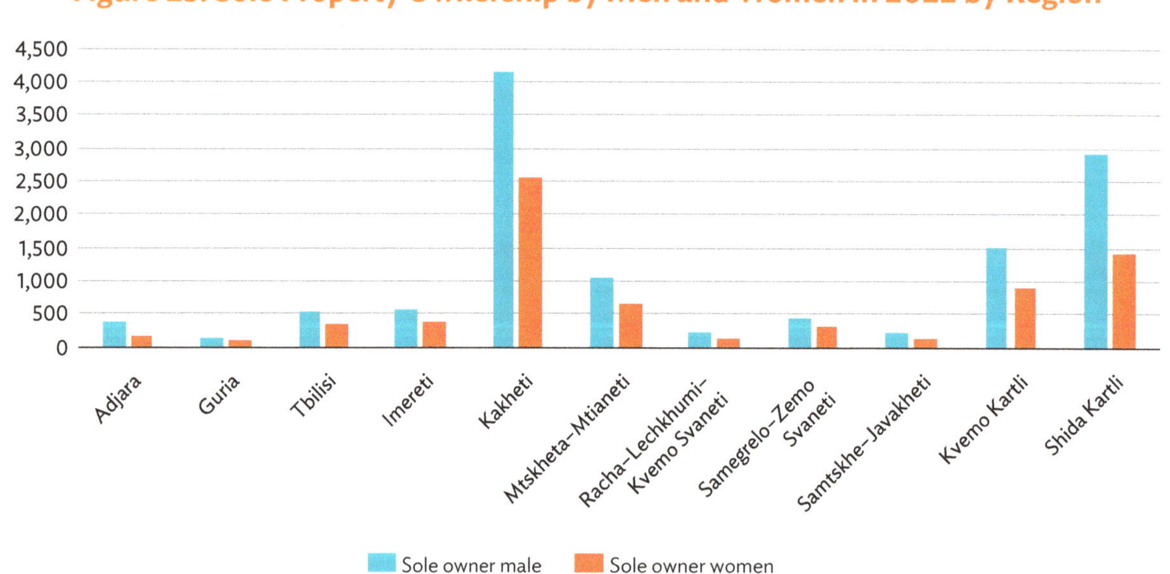

Source: Authors' elaboration based on studies.

The gender wage gap plays an important role in contributing to the significant disparity between properties owned by men and women (Figure 23). According to national statistics for 2022, 12,053 properties were solely owned by men, while the corresponding figure for women was almost half at 6,097. This stark contrast underlines the impact of unequal earnings on property ownership and highlights the need to address gender-based economic disparities to promote greater equity in real estate ownership.

Figure 23: Sole Property Ownership by Men and Women in 2022 by Region

Source: National Agency of Public Registry.

Homeownership

Another important factor to consider to better understand the national context is the homeownership rate. Georgia currently has a homeownership rate of 92%, one of the highest in the Commonwealth of Independent States and significantly higher than the EU-27 average of 69.7% (footnote 21). The lowest homeownership rate is in Tbilisi, which is due to higher prices driven by foreign investor demand, although it is still well above the average for other countries, as can be seen in Figure 24.

Figure 24: Homeownership Rates

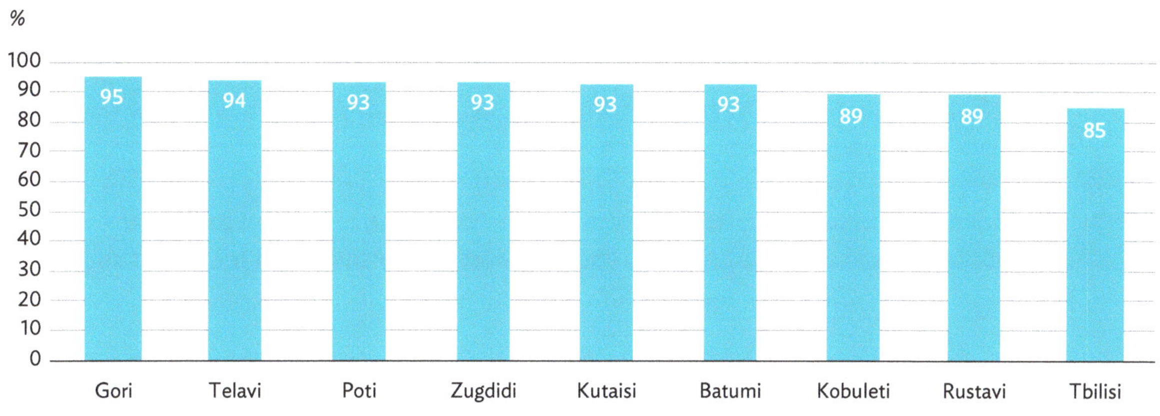

Source: Global Property Guide.

Conclusion and Recommendations

In conclusion, while a 5% increase in house prices could potentially deepen inequalities between different household income groups, the assessment does not provide any evidence or findings to suggest that this scenario would lead to significant negative consequences in the specific context of Georgia. This is primarily because the proportion of income currently spent on mortgage loans is well below the thresholds set by national regulation, which are considered a prudent share of expenditure, and a 5% increase in house prices is unlikely to change this.

This is emphasized by the market dynamics and context in Georgia. Spending on real estate by low-income groups is primarily aimed at securing basic essential housing, to which the majority of the Georgian population has access (more than 90%).

However, it is worth noting that despite these positive indicators in terms of housing accessibility, Georgia still grapples with serious issues of poverty and economic challenges, coupled with important gender issues, especially for low- and middle-income groups.

Therefore, it is important that the government, in collaboration with other public and private organizations, proactively plans and implements measures to ensure that as many households as possible have the opportunity to benefit from newly constructed energy-efficient buildings and that gender gaps are reduced. Such recommendations include the following:

(i)   **Targeted subsidies or grants** specifically designed to support low- and middle-income households and women in vulnerable economic situations. These subsidies could aim to mitigate the potential impact of the 5% increase in house prices and ensure affordability and access to essential housing.

(ii)   **Benefits for households acquiring new energy-efficient houses.** Benefits for low-income households purchasing this type of property, including special programs for women, could include, for example, a tax exemption for a certain number of years. This would not only mitigate the negative effects of rising prices, but also promote the development and growth of the sector.

(iii)   **Affordable housing initiatives in partnership with developers** with a focus on energy efficiency. The government could work with developers to create housing projects that meet the needs of lower-income groups, including specific programs for women to provide them with sustainable and cost-effective housing options. For example, the government could offer fiscal benefits to constructors of such houses.

(iv)   **Transformative changes to address the gender wage gap.** This is an issue that goes beyond the variable examined here. It requires a comprehensive set of measures, such as gender-inclusive policies, awareness campaigns, strong regulations for equal pay, etc.

(v)   **Public–private partnerships.** Encourage collaboration between the government and the private sector to create innovative financing models that support housing affordability for different income groups.

(d)   Variable 4: Distributional Impact on Renting Affordability

Summary of Model Outputs

A rise in housing prices is likely to be reflected in a rise in rents for these houses. Such an increase may be a problem for lower-income groups seeking rental accommodation.

However, the impact of this problem in Georgia is somewhat mitigated by the country's high homeownership rate (+90%), which suggests that the vast majority of the population already owns a residence and has no need to rent. In fact, renting is mainly associated with foreigners coming to work in the big cities, as well as people moving from rural areas to the big cities to find work.

Although this is not a significant problem for the majority of the population, the impact on the distribution of rent is assessed, particularly as from a just transition perspective, the ability of people from low-income groups to move for better job opportunities may be affected.

As can be seen in Figure 25, the increase in rental prices will be even more challenging for the lowest-income groups (groups 2–7) as they are already grappling with a financial burden, having to spend more than 45% of their income on rent. For these vulnerable groups, a 5% increase in rent prices would represent an additional burden that could increase the current share of income spent on rent by up to 3.9%. For the other household income groups, the increase in their current share of income would be less than 1.8%.

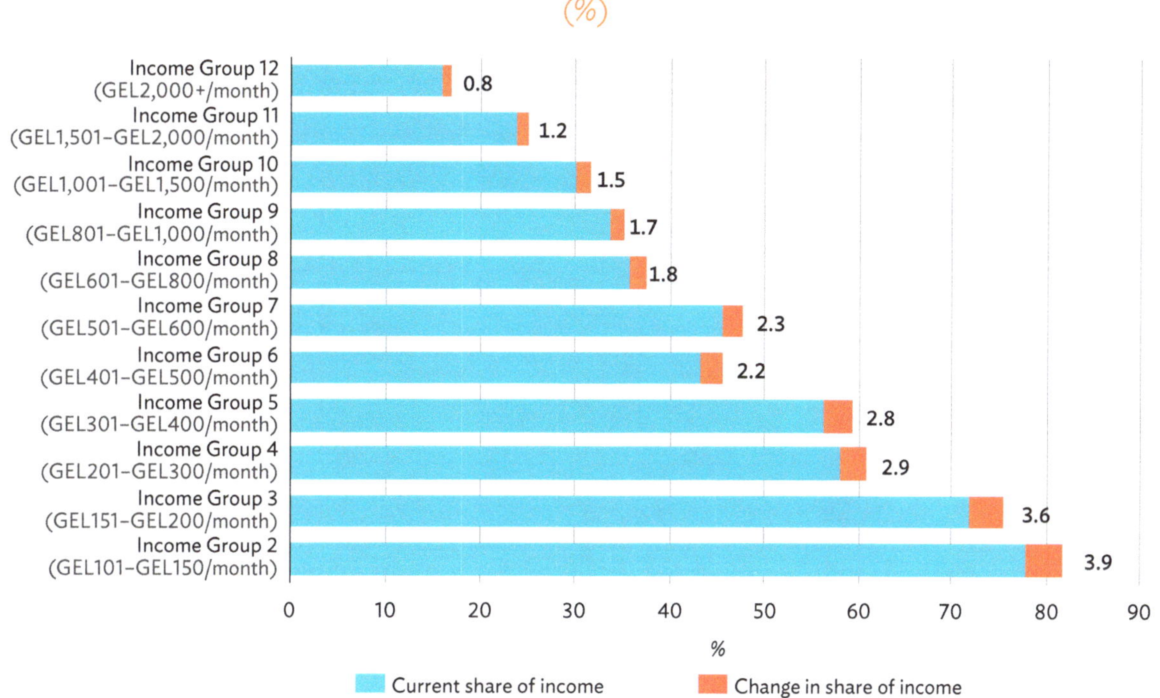

Figure 25: Change in the Share of Rent Expenditure by Income Group (%)

Source: Authors' elaboration based on data from household surveys and assumptions.

It is important to consider that the number of households in these vulnerable categories represent almost a quarter of the total number of households in Georgia as seen previously in Figure 8.

In terms of the nominal increase, this means an increase of GEL5–GEL10 per month for income groups 2–7 and GEL10–GEL25 per month for income groups 6–12 (Figure 26).

Figure 26: Nominal Increase per Month by Income Group (GEL)

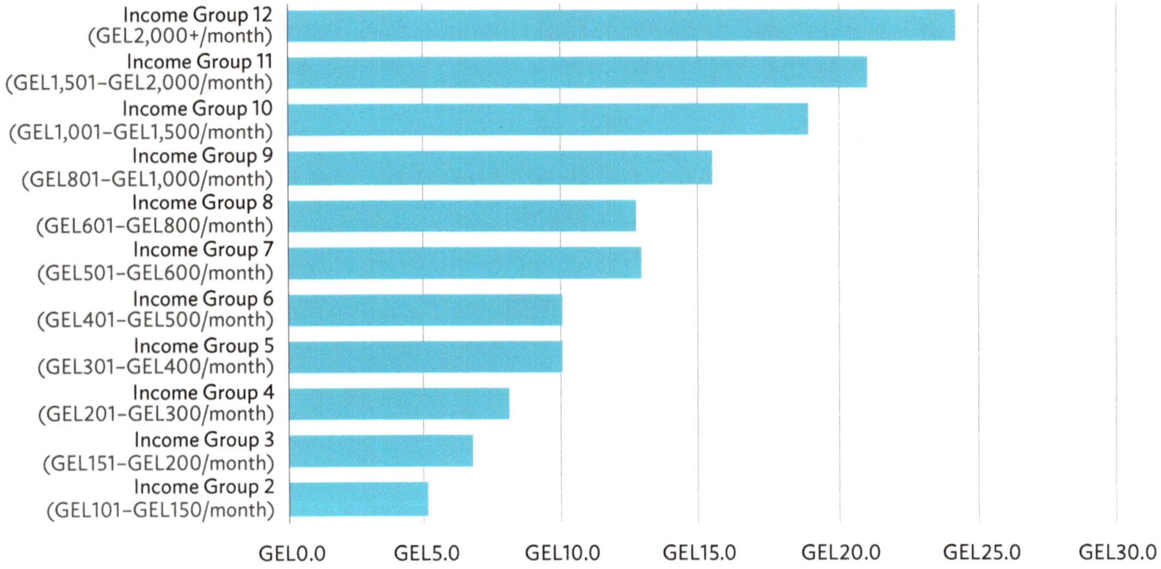

Source: Authors' elaboration based on data from household surveys and assumptions.

Gender perspective

Following the same logic mentioned in the variable assessment above, rent prices may also disproportionately affect women as a result of the existing gender wage gap (Figure 27).

Figure 27: Change in the Share of Income Spent on Rent by Men Compared to Women

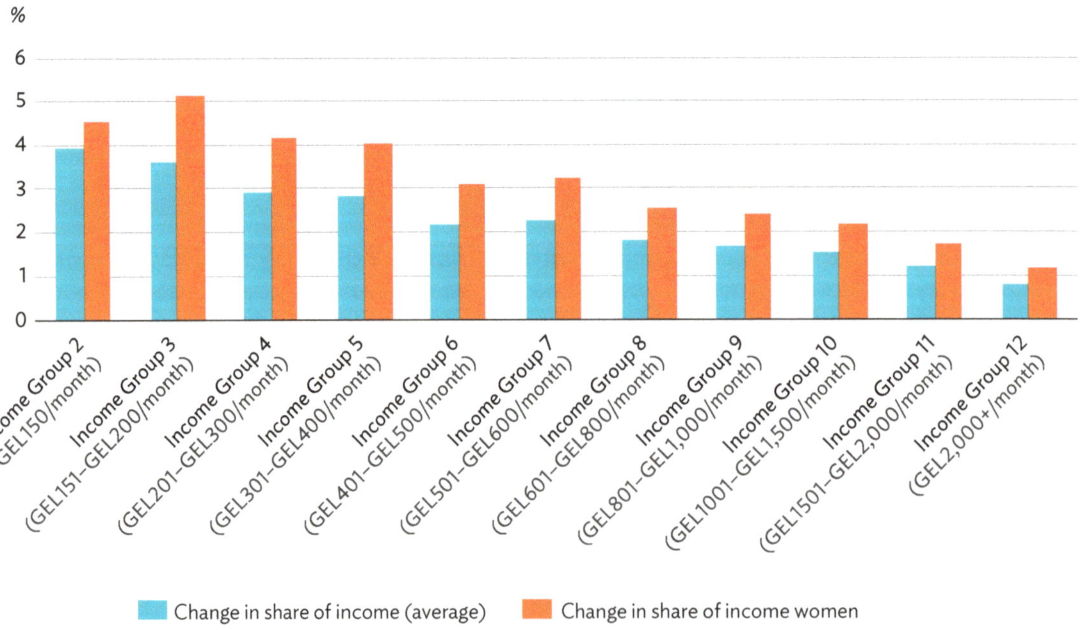

Source: Authors' elaboration based on data from household surveys and assumptions.

Conclusion and Recommendations

The model outputs show that a 5% increase in rental prices for newly built energy-efficient homes is likely to lead to a significant increase in the proportion of income currently spent on rent by the lowest-income groups (>2%). This increase threatens to make it more difficult for the lowest-income groups to access such houses and exacerbate inequalities between high- and low-income groups.

As mentioned above, while renting is not common in Georgia due to the high rate of homeownership, this could pose a challenge for people moving to major cities in search of better job opportunities. It is therefore important that the government, in collaboration with other organizations, proactively plans and implements measures that both mitigate the negative impacts and maximize the potential opportunities. Such recommendations include the following:

(i) **Rent subsidies tailored for most vulnerable groups.** These subsidies can help to alleviate the increased burden on their income caused by rising rents and ensure access to housing.

(ii) **Rent control measures in collaboration with property owners and developers** to implement sustainable rent control measures. These may involve granting fiscal benefits to property owners in return for setting a predefined affordable rent for newly built energy-efficient homes. Such a measure not only has an impact on low-income households seeking an affordable rent, but also stimulates demand for the construction of such buildings and thus economic growth in this sector. An example of this is the PINEL law in France, as seen in Box 1.

(iii) **Public–private partnership** between government and the private sector to develop innovative solutions for affordable and energy-efficient rental housing. Joint efforts can bring together expertise and resources to effectively address housing challenges.

By proactively implementing these recommendations, the government can not only mitigate the negative impact of rising rental prices, but also create a more inclusive and resilient housing environment that promotes access to energy-efficient housing for a larger portion of the population.

Box 1: PINEL Law in France

France's PINEL law, introduced in 2014 and extended until 2024, is a notable initiative to promote the construction of affordable rental housing. This law provides tax incentives to property owners who commit to renting out their newly constructed or rehabilitated properties at a predetermined affordable rent for a specified period of 6 to 12 years. The incentives vary depending on the rental period chosen. The PINEL law is designed to address the shortage of affordable rental properties, particularly in high-demand urban areas. By encouraging private investment in the rental market, the law seeks to increase the supply of affordable housing options and thus promote a more inclusive and dynamic housing market in France.

Source: Ministry of the Economy, Finance and Industrial and Digital Sovereignty. 2024. Investissement locatif: tout savoir sur la réduction d'impôt "Pinel".

(e)   Variable 5: Distributional Impact on Energy Savings

Summary of Model Outputs

The proposed reform can improve the well-being of households by promoting energy-efficient buildings, leading to significant energy savings. Although the current scenario in Georgia does not present energy expenditure as a pressing concern given the price control measures taken by the government, it is important to anticipate potential challenges in the near future when the market moves to an open structure. The introduction of an open market may lead to fluctuations in energy prices, making energy efficiency a key factor for sustainable and cost-effective living, especially for the lowest-income households.

In light of this, the variable aims to assess the extent to which different household income groups can benefit from the potential energy savings of such a reform.

In the absence of a primary estimate for the average energy savings from the reform, this assessment is based on a hypothetical scenario in which the newly introduced energy-efficient buildings result in a 20% reduction in current electricity bills.

Figure 28 shows the potential change in the share of electricity costs for each household income group.

Figure 28: Change in the Share of Expenditure on Electricity by Income Group (%)

Income Group	Current share of income	New share of income
Income Group 12 (GEL2,000+/month)	1.5	1.2
Income Group 11 (GEL1,501–GEL2,000/month)	2.0	1.6
Income Group 10 (GEL1,001–GEL1,500/month)	2.5	2.0
Income Group 9 (GEL801–GEL1,000/month)	2.8	2.3
Income Group 8 (GEL601–GEL800/month)	3.2	2.5
Income Group 7 (GEL501–GEL600/month)	4.2	3.3
Income Group 6 (GEL401–GEL500/month)	4.2	3.4
Income Group 5 (GEL301–GEL400/month)	4.4	3.5
Income Group 4 (GEL201–GEL300/month)	6.6	5.3
Income Group 3 (GEL151–GEL200/month)	8.8	7.0
Income Group 2 (GEL101–GEL150/month)	11.0	8.8

Source: Authors' elaboration based on data from household surveys and assumptions.

As observed, the biggest beneficiaries of these energy savings would be the lowest-income groups. Within this demographic, the share of income currently spent on electricity costs can be reduced by up to 2.2%. It should also be noted that this effect could be even greater if electricity prices rise following the implementation of the open market reform. The resulting surplus in household finances could mean a significant boost in purchasing power for these households, enabling them to allocate funds to other important areas of the economy.

Gender perspective

Energy savings also offer the opportunity to narrow the gender gap by facilitating women's access to such energy-efficient buildings. As shown in Figure 29, which considers the national gender wage gap, women-owned households in the lowest-income groups could experience a notable reduction of up to 3.2% in the proportion of income spent on electricity expenses. This represents a 42% greater reduction than the average and highlights the potential of energy-efficient reforms to disproportionately benefit women in economically vulnerable situations.

Figure 29: Change in the Share of Electricity Expenditure by Income Group and by Gender (%)

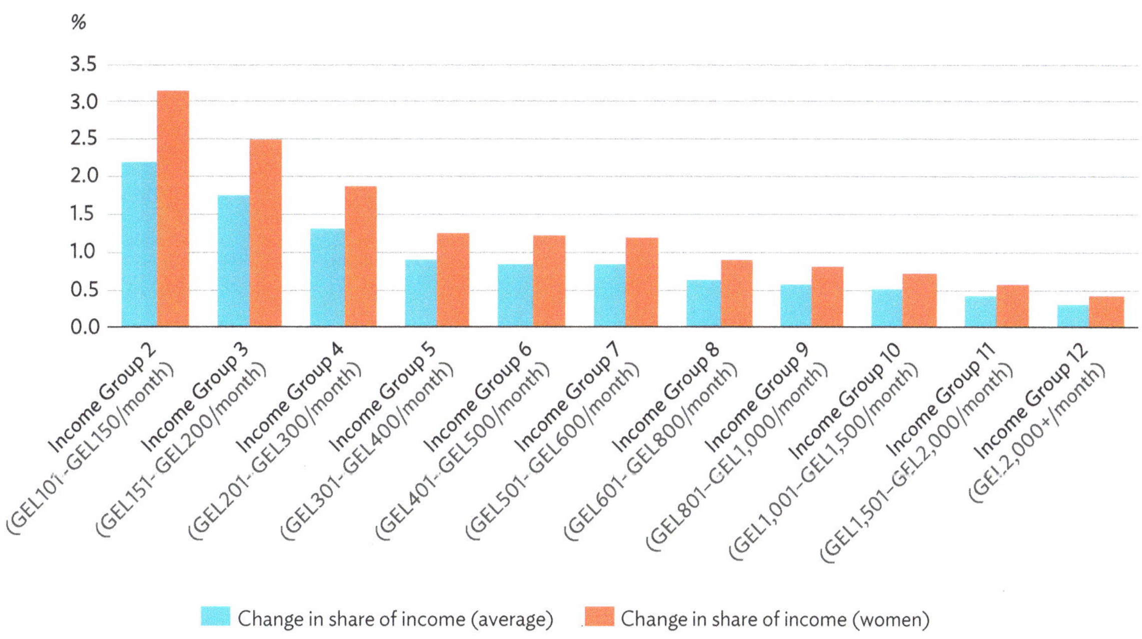

Source: Authors' elaboration based on data from household surveys and assumptions.

Conclusion and Recommendations

The evaluation of this variable underscores that the potential benefits of energy savings are most pronounced for the lowest-income groups, as well as vulnerable groups such as low-income women, provided they have access to these new energy-efficient buildings.

To ensure that all segments of the population benefit from this advantage, it is essential that the recommendations listed in variables 3 and 4 are implemented. These measures aim to foster both the purchase and rental of these homes so that they are accessible to a broad spectrum of the population and not just higher-income groups. This approach is in line with the overarching goal of equitably distributing the benefits of energy efficiency across different socioeconomic segments.

(f)   Recommendations on Activity 2

Table 12 summarizes the key recommendations for each measure based on the identified impacts and the corresponding significance assessment. Appendix 2 contains details on the significance assessment.

Table 12: Summary of Findings, Recommendations, and Impact Significance Assessment

Variable	Type	Significance	Impact Summary	Recommended Measure
Variable 1: Distributional impact on constructors' suppliers	RISK	MODERATE	5,324 MSMEs potentially at risk, mainly due to the need to invest in new machinery (~ $80,000)	• Offer preferential financing and de-risking programs. • Stimulate local demand through financial incentives for constructors and public procurement policies. • Capacity building and training programs
Variable 2: Distributional impact on constructors	RISK	NEGLIGIBLE	No significant impact estimated	Monitoring
Variable 3: Distributional impact on house acquisition affordability	RISK	MINOR	No significant impact on access to housing due to high ownership rates and low debt ratio.	• Develop energy efficiency programs to promote access to energy-efficient buildings for low-income groups to fight economic challenges faced by households: targeted subsidies, fiscal benefits for households that purchase energy-efficient buildings, fiscal benefits for constructors that sell to low-income households at a lower price.
Variable 4: Distributional impact on renting affordability	RISK	MINOR	No significant impact on renting due to high ownership rates. Could have some moderate impact on the lowest-income groups already struggling with high rents (people in the bottom 25% of the total population) - particularly women-led households.	• Develop energy efficiency programs to promote access to energy-efficient buildings for low-income groups, especially vulnerable women-led households, to fight economic challenges faced by households: targeted rent subsidies, rent control measures in collaboration with property owners.

continued on next page

Table 12 *continued*

Variable	Type	Significance	Impact Summary	Recommended Measure
Variable 5: Distributional impact on energy savings	OPPORTUNITY	MODERATE	Could deliver significant savings for the lowest-income groups (people in the bottom 25% of the total population, especially women-led households.) – ~2.2% reduction in the current share of income spent on energy expenses	Implement the measures listed in variables 3 and 4 to facilitate access to these buildings for low-income groups and vulnerable women-led households.

MSMEs = micro, small and medium-sized enterprises.
Source: Author's own elaboration based on studies.

The adoption of the energy efficiency reform for the construction of new buildings presents a mixed bag of opportunities and risks. It has the potential to promote both economic development and social welfare, but also poses challenges for certain segments of the population. To make this transition effective, it is crucial to carefully consider the potential impact on various stakeholders and implement targeted measures to mitigate any adverse effects.

Upstream, the main significant impact is the risk to MSMEs supplying construction materials. This is primarily due to the potential need for these businesses to invest in new machinery and equipment to comply with the new standards. While this challenge may seem daunting at first, it also presents an opportunity to promote the growth of an underdeveloped sector in the country. Tailored financing programs, capacity-building initiatives, and stimulation of local demand should be implemented to overcome this challenge.

Downstream, the most significant impact is the potential savings in energy consumption, especially for the lowest-income households who spend a higher share of their income on this expenditure. On the other hand, contrary to initial concerns, the impact on renting and house acquisition is low or negligible due to the socioeconomic circumstances in Georgia (high ownership rates and safe debt ratios).

However, these positive effects depend on access to energy-efficient buildings. The main challenge is therefore to ensure that low-income households have access to these new energy-efficient buildings so that they too can benefit from the potential savings in energy consumption.

The recommended measures focus first on ensuring that the reform is socially accepted and well received by the market. To achieve this, it is recommended to focus on awareness-raising campaigns to highlight the potential benefits of the reform to the different segments of the population.

Once the market is established, the authorities should promote the implementation of energy efficiency programs for low-income households: offer financial incentives to low- and middle-income households to access this type of property. This could involve fiscal benefits, tax breaks, or subsidies for households that purchase or rent energy-efficient buildings. Following the example of some European countries, Georgia can take measures to provide fiscal benefits and advantages to households that purchase new energy-efficient buildings and rent them at an affordable price for low- and middle-income households. These types of measures aim to provide benefits to both the buyer and the renter and expand access to energy-efficient housing.

As Georgia goes through this transformative phase, the success of the reform lies not only in mitigating the negative impacts, but also in the equitable distribution of benefits. The government's commitment to implementing these recommendations will determine the degree of inclusivity achieved and set the stage for a sustainable and just transition.

Activity 3:   Intercity Bus Passenger Transport Reform

1.   Scope of Activity

To reduce greenhouse gas (GHG) emissions in the transport sector, the government is planning to implement an intercity passenger transport reform that will create a new legal framework for the sector. The aim of the reform is to improve the quality of intercity passenger transportation services by creating a competitive environment and improving safety levels and environmental aspects, integrate the country's road transport system into a modern framework, promote tourism, and create employment opportunities for qualified professionals in passenger transportation. A key focus is on addressing the disorganized bus service, which operates from overcrowded assembly points rather than designated bus stations, leading to increasing traffic congestion and environmental degradation. The reform aims to curb the practices of unregistered transport companies that operate without a timetable and violate traffic rules, environmental norms, and drivers' working hours. It also targets the increase of unskilled workers using substandard vehicles for passenger transportation, with a focus on raising technical specifications and comfort standards to meet overall environmental conditions. The impact of the following amendments are assessed in this report (Table 13).

Table 13: Amendments to Current Regulations in Georgia

Amendment	Description
Amendment to the Law of Georgia "On Motor Transport"	Definition of permitted activities: implementation of a new type of permit. Increase carrier accountability toward passengers: adapt passenger transportation to modern requirements. List conditions and required documents to be submitted. Rules and conditions for bus station certification together with specific requirements: including the criteria for the term, issuance, refusal, suspension, renewal, and cancellation of bus station certificates.
Amendments to the Code of Administrative Offenses	Persons who violate the regulations are subject to an initial fine of GEL1,000. For repeated offenses, the fine increases to GEL2,000. The law stipulates that violations of the technical regulations for the operation of bus stations can be punished with fines of up to GEL5,000.
Amendment to the "Local Self-Government Code" in the Organic Law of Georgia	Strengthen the authority of municipalities by introducing new powers to ● make decisions on the suitability of establishing a bus station within the municipality's territory; and ● establish a list of settlements where it is not obligatory to board transportation solely at the bus station if there is no bus station within the administrative boundaries of these settlements.
Amendments to "Licenses and Permits"	Include a permit for motorized passenger transportation.
Amendments "On Control of Entrepreneurial Activity"	The Land Transport Agency to conduct business activity inspections without the need for a court order.
Environmental improvements	The alignment of these reforms with European transport requirements is expected to lead to environmental improvements. Compliance with international standards and regulations is likely to lead to the adoption of cleaner and more sustainable technologies in passenger transportation. Improved road safety measures combined with stringent adherence to environmental protection norms are likely to mitigate the adverse ecological impact of disorderly bus services and unregulated transportation practices. In addition, the focus on improving service quality for customers is in line with the global trend toward more sustainable and environmentally conscious travel modes and encourages the adoption of eco-friendly practices and vehicles. Ultimately, these reforms aim not only to improve the efficiency and safety of passenger transportation, but also to make a positive contribution to the broader global goal of sustainable and environmentally responsible mobility.

Source: Consultations with government stakeholders.

2.   Value Chain Map

Figure 30 shows the general value chain for the intercity bus passenger transport reform.

Figure 30: Value Chain Map of the Intercity Bus Passenger Transport Reform

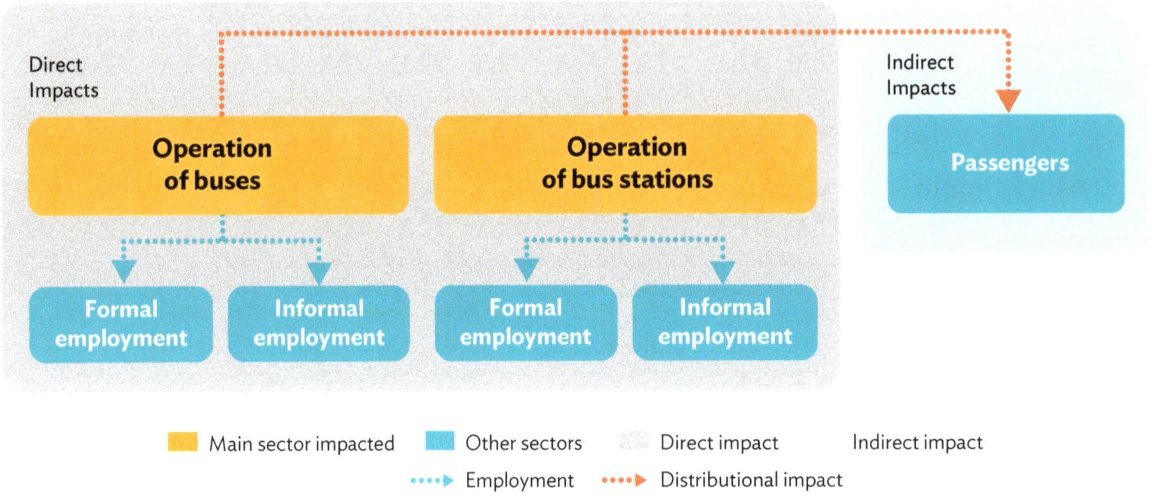

Source: Author's own elaboration.

3.   Qualitative Assessment: Identification of Affected Stakeholders, Risks, and Opportunities

The main stakeholders affected in the transport sector are likely to be the operators of buses and bus stations, especially informal workers, who will have to abide by the new compliance requirements for obtaining permits. Informal workers make up a large share of the market and the reform, even with the planned transition period, may affect their capacity to continue operations if they do not have the means to comply with the requirements.

Another large group that will most likely be affected are the passengers who rely on the existing Georgian transportation system. Between 30% and 80% of passengers rely on these routes. The regulation of bus operators will most likely affect fares for passengers who may have relied heavily on the services of informal bus operators. These passengers may also be heavily dependent on the informal bus stops to access the bus service. However, with the redesign of bus routes, bus stops and bus stations may be removed as part of the government's efforts to streamline the system, which would affect passenger mobility, especially in remote areas.

The removal of existing bus stops and bus stations may also impact the local informal workers who have established businesses in these areas. As their customer base is largely made up of bus operators, bus station workers, and passengers, the potential removal of these bus stops or bus stations may affect their livelihoods.

4.   Socioeconomic Variables Analyzed

A quantitative model was defined, identifying three main variables potentially affected by the new regulations (Table 14). The assessment aims to measure the impact from a just transition perspective, focusing on population groups that are more exposed and vulnerable to ticket price oscillations, job market changes, and changes in service characteristics. The methodology was developed using the available data obtained through desk research and in-country consultations. The defined scenarios were adapted to the Georgian population and socioeconomic characteristics.

Table 14: Quantified Impacts of the Intercity Transport Reform

	Socioeconomic Variable	Type of Impact	Quantification Approach
1	Distributional impact on service users	Indirect	The impact study has two objectives: first, to quantify the impact of ticket price changes on users using price scenarios and demand elasticities; and second, to assess the impact of changes in the use of the service. In both cases, the distributional impacts are analyzed by population groups affected by socioeconomic characteristics such as income level, gender, and age groups.
2	Impact on compliance costs for service providers	Direct	Estimates of the upfront costs that all drivers will have to pay to adapt to the new regulations. One-time costs and annual costs are part of the analysis. Recommendations are made on how the financial burden can be mitigated. It also considers the impact on informal workers who will face additional challenges to comply with the new regulations.
3	Impact on local communities in the station surroundings	Indirect	Estimates of the number of street vendors and other informal workers who depend on the activities around the bus stations and could be affected by the reform. The recommendations target women and other vulnerable minorities.

Source: Author's own elaboration.

5.   Model Outputs and Recommendations

(a)   Context Analysis

Socioeconomic Profile of Intercity Transport Users

Intercity transportation in Georgia is dominated by private service providers made up of formal and informal drivers who determine the service fare, the frequency, and therefore the quality of travel. The bus stations are also privately managed. The owners levy a discretionary fee on drivers and at the same time influence the dynamics around the stations. In terms of gender gap, the transport sector is generally dominated by men, who make up 79% of the total workforce.[23]

---

[23]   Ministry of Economy and Sustainable Development (MOESD). 2021. Survey on the Current Demand and Future Intentions for Labor and Skills in the Transport and Energy Sectors.

On the demand side, 15% of respondents to the Household Income and Expenditure Survey[24] reported allocating part of their expenses to intercity transport services (Figure 31).

Figure 31: Participation of Intercity Bus Transport Users in Total Respondents

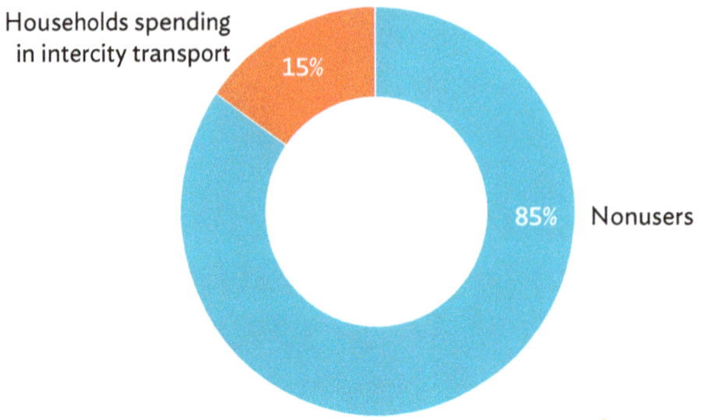

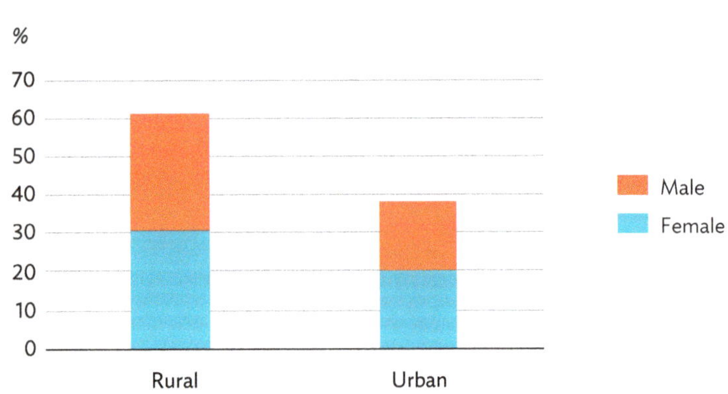

Source: National Statistics Office of Georgia. Geostat Database. https://www.geostat.ge/en (accessed 5 March 2024).

Of those using the services, more than 60% of households are located in rural areas, with the proportion of women among household members being slightly higher at 52% (Figure 32). Local consultations with gender specialists, CSOs, and other relevant stakeholders revealed that women rely heavily on intercity transport for their essential mobility needs. Women use bus services to get to work, take their children to school, and shop for groceries, to name a few. Only a small percentage of women own a private car, 10.4%.[25] When families share a car, it is usually the man who drives due to his own mobility needs. In this context, improving the service quality, correcting inaccurate schedules, and creating a more robust behavioral framework for users of the service would contribute to greater safety and expanded mobility options for women.

Figure 32: Gender Distribution of Users by Region

Source: National Statistics Office of Georgia. Geostat Database. https://www.geostat.ge/en.

---

24   Geostat. 2022. Household Income and Expenditure Survey.
25   A. Iluridze and T. Verulashvili. 2021. *Women and Mobility: Gender Aspects of the Women Daily Movement.* Public Defender (Ombudsman) of Georgia and GIZ. https://ombudsman.ge/res/docs/2021092012383614913.pdf.

Other vulnerable groups would also be affected by the changes to the service. Statistics show that among users, 24% of household members are under 18 years old and 16% are over 65 years old (Figure 33). They could also be more affected by the impact of the reform, as they generally have little access to private cars and less economic independence and solvency.

Figure 33: Age Distribution of Intercity Bus Transport Users

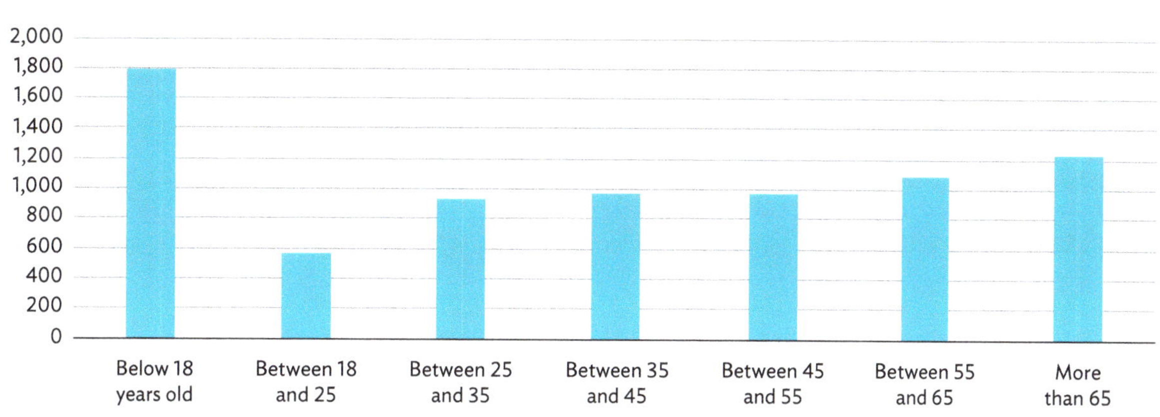

Source: National Statistics Office of Georgia. Geostat Database. https://www.geostat.ge/en (accessed 5 March 2024).

Another relevant group is the student population. About 60% of households have members with a student in secondary or vocational education, who tend to rely heavily on the bus service (Figure 34). They should also be more affected by ticket price changes and improvements in service quality.

Figure 34: Education Level of Intercity Transport Users

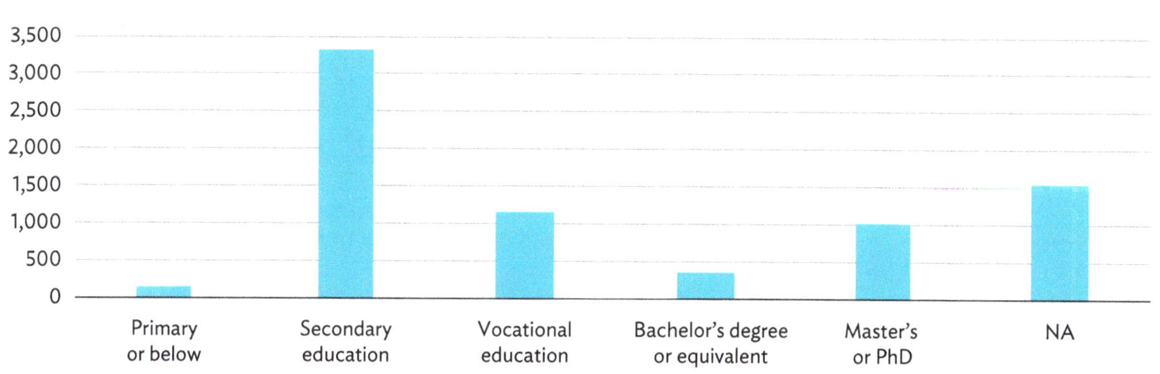

NA = not applicable.
Source: National Statistics Office of Georgia. Geostat Database. https://www.geostat.ge/en (accessed 5 March 2024).

Vulnerable users also include workers who travel to remote areas for work, people with disabilities, internally displaced people, and those who rely on the intercity transport network to commute between remote areas and health-care facilities.

Participation of Different Groups in Intercity Transport

Figure 35 shows the contribution of each income percentile to the total declared intercity transport expenses. The average income of each percentile is shown on the left vertical axis, with the first two percentiles below the Georgian mean salary of GEL1,040 per month calculated by Geostat for all economic sectors.[26]

There is a positive correlation between the average income per group and their contribution to the total intercity transport expenses: Almost 27% of total intercity transport expenses are driven by lower income groups of the population, which confirms that they are the most exposed to both positive and negative changes.

Figure 35: Average Income Distribution of Intercity Household Users and Their Share of Total Intercity Expenses
(GEL)

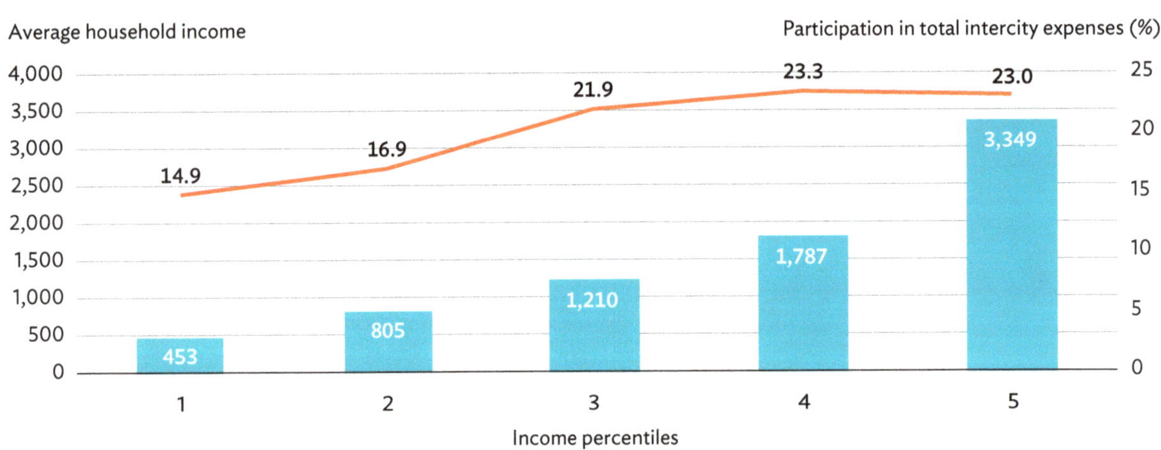

ᵃ Average monthly household income: Percentile 1: GEL452.5; Percentile 2: GEL805.3; Percentile 3: GEL1,210.2; Percentile 3: GEL1,787.0; Percentile 4: GEL3,349.0.
Source: National Statistics Office of Georgia. Geostat Database. https://www.geostat.ge/en (accessed 5 March 2024).

---

[26] Geostat. 2022. Median Earnings.

Gender Dimension

The gender dimension in access to intercity transport in Georgia is also relevant. In Georgia, gender roles are synonymous with unequal labor opportunities for women. This issue is reflected in their purpose of travel, distance, and choice of transportation and means. Women travel shorter distances and more frequently than men, with the frequency increasing during peak hours. Women's main purpose of travel is usually to accompany children and older people.[27]

Women's mobility is directly related to their income: 41.3% of women with an income of less than GEL300 travel daily, rising to 70.3% in the income groups between GEL300 and GEL600 and to 75% for those with an income of more than GEL600, as shown in Figure 36.

Figure 36: Women's Movement and Income

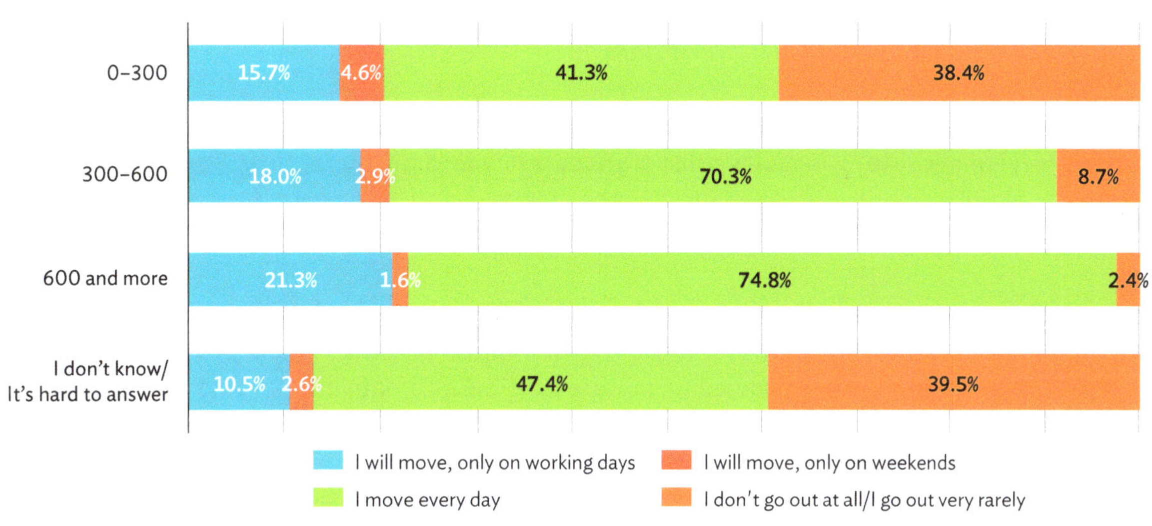

Source: A. Iluridze and T. Verulashvili. 2021. *Women and Mobility: Gender Aspects of the Women Daily Movement*. Public Defender (Ombudsman) of Georgia and GIZ. https://ombudsman.ge/res/docs/2021092012383614913.pdf.

The safety of women in public transport is also an important issue. About 62% of female students have experienced some form of sexual harassment in public spaces, of which 33% are frequently observed on public transportation (footnote 27). This is a restriction on women's free mobility that affects their right to education and economic and social participation. Implementing public policies to improve inclusiveness and safety conditions for women on public transportation will not only strengthen public cohesion, but also promote women's participation in society and improve their educational and employment opportunities.

---

[27] A. Iluridze and T. Verulashvili. 2021. Women and Mobility: Gender Aspects of the Women Daily Movement. Public Defender (Ombudsman) of Georgia and GIZ. https://ombudsman.ge/res/docs/2021092012383614913.pdf.

Low-income women in particular, who live on the periphery, are most dependent on intercity public transport. They are less likely to own a private vehicle and less likely to have a driver's license. They rely on public transport and its timetables to get home earlier and may miss out on employment and educational opportunities because they do not have access to mobility (footnote 27).

(b)   Variable 1: Distributional Impact on Service Users

The purpose of this section is to identify the differential impact on service users and present a set of recommendations for the government to mitigate the response to ticket prices and reach a broader spectrum of the Georgian population with the improvement of service quality.

Summary of Model Outputs

Distributional impact of ticket price variations

The current market structure, which is dominated by private drivers and involves many informal service providers, impairs price transparency and predictability. Based on the experiences of other countries documented in the literature and the urban transport reform implemented in Tbilisi, there are reasons to expect an increase in fares.

However, the purpose of this assessment is not to determine whether or not there will be a price increase, but rather to examine what would happen if there were a price increase. In this line, three price scenarios were modeled to capture possible alternatives. It should be noted that the final impact on market prices depends to a large extent on the discretion of the service providers, as government authorities do not intend to intervene in the ticket fare.

To capture the expected reaction of transport users to price changes, a literature review was conducted to find estimates of the elasticity of demand in public transport. A number of studies have shown that demand tends to react negatively to price changes. The estimated elasticity of demand for bus transportation is −0.56, which means that a 1% increase in fare leads to a 0.56% decrease in demand.[28] It is worth noting that a different response can be expected depending on income level, car ownership, distance traveled, and time period in which the service is used (Figure 37).

[28]   Y. Kholodov et al. 2021. Public Transport Fare Elasticities from Smart Data: Evidence from a Natural Experiment. *Transport Policy.* 105. pp. 35–43.

Figure 37: Elasticity Estimates from a Number of Studies

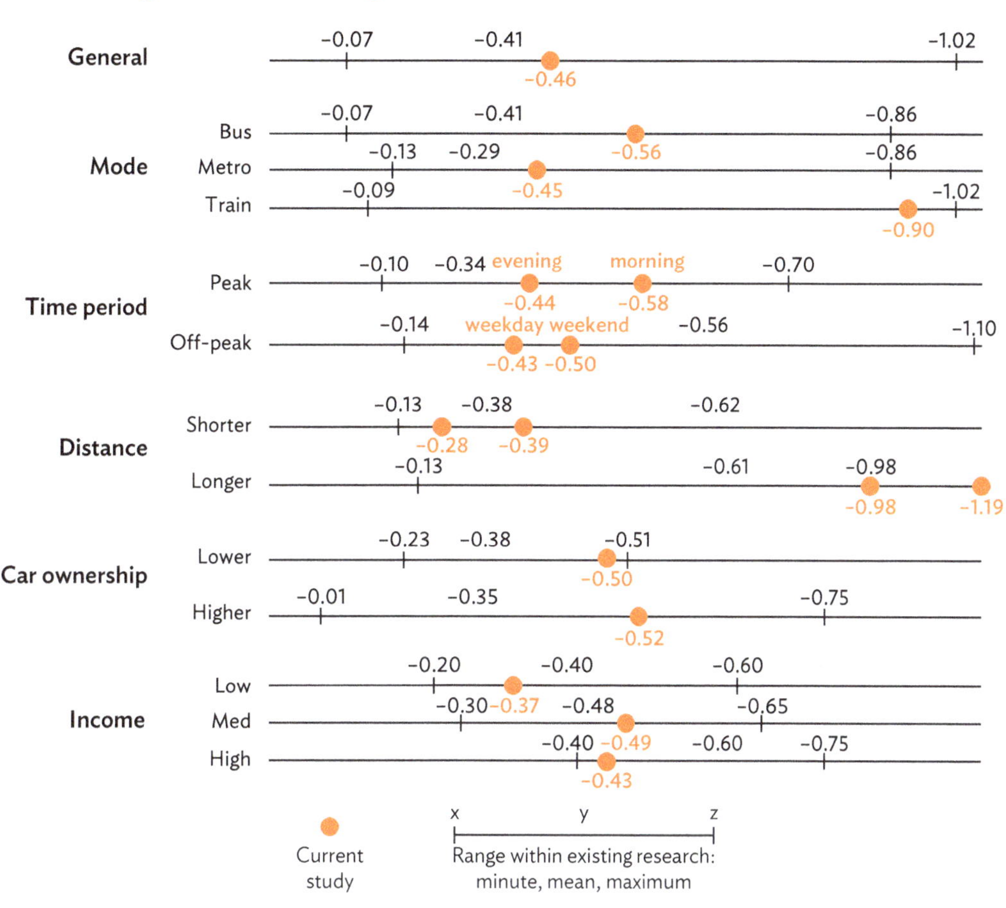

Source: Y. Kholodov et al. 2021. Public Transport Fare Elasticities from Smart Data: Evidence from a Natural Experiment. *Transport Policy*. 105. pp. 35–43.

The elasticity factor considered in this assessment is 0.56, based on the elasticity of bus transportation shown in Figure 37. Demand in the various price increase scenarios would then be affected as shown in Table 15.

Table 15: How Demand in Price Increases Will Be Affected

Price Scenario	Expected Effect on Demand
S1: Price increase = 5%	Entails a **2.8% decrease** in the overall bus demand level
S2: Price increase = 10%	Entails a **5.6% decrease** in the overall bus demand level
S3: Price increase = 15%	Entails an **8.4% decrease** in the overall bus demand level

Source: Author's elaboration based on Figure 37.

Acknowledging that the demand elasticity also varies depending on other characteristics, further socioeconomic relevant categories were introduced to capture the heterogeneous effects of ticket price variations.

- **Income level: Distinguishing between two distinct groups—low- and low-middle income, and upper-middle and high-income—revealed compelling insights.**

  - ☐ **Low-income users proved to be the most inelastic group** with a demand elasticity of –0.37. This implies that they are less likely to change their demand behavior significantly in response to price changes, as they use this service for essential activities such as work and study. Consequently, higher ticket prices may affect their purchasing power for other essential expenses.

  - ☐ **Middle-income users, on the other hand, generally exhibit greater elasticity, with an elasticity of –0.49.** Their access to alternative modes of transportation contributes to this elasticity. However, it is worth noting that in the current Georgian context, middle- and high-income groups hardly use public transportation for intercity commute. This is not due to high prices, but rather problems related to service such as safety and reliability. Fluctuations in ticket prices therefore appear to be inconsequential for this demographic, while improvements in other factors could potentially increase the use of this service— even if this entails a higher ticket price.

- **Frequent user groups were also defined, including** students, the youth, the older people, and people with disabilities.

- **The rural population was also assessed as a standalone group,** as low-income workers in rural areas often rely on intercity transport to get to remote areas where they work.

- **Low- and middle-income women were disaggregated to account for gender disparities in income distribution and access to services in the Georgian context.** Women's mobility is directly tied to their income: 38.4% of very low-income women rarely or never leave their homes because they do not have access to mobility services.

Distributional impact on service quality and use

One of the expected outcomes of the reform is the improvement of the quality of intercity transportation for different groups of Georgian society. The expected outcomes of the reform are distributed among different household income groups (Table 16).

Table 16: How the Current Issues Could Be Improved with Reform

Population Group	Current Use	Current Quality Issues	Expected Outcomes of the Reform
Low-income groups	Currently use the service as they have no other alternative	Users suffer from issues related to safety, reliability and efficiency, especially vulnerable groups such as women.	Users will benefit from a safer, more reliable and more efficient service. It is important that poorly connected areas are mapped and the respective service provision are regulated to provide a safety and reliable service for workers to accomplish their working schedules.
Middle- and high-income groups	Barely use the service	Concerns about reliability, efficiency, safety and transparency.	Attract more passengers through a more frequent, reliable, comfortable, and affordable intercity bus service.

Source: Author's own elaboration.

Assessment of the use of services by gender and vulnerable groups

As mentioned earlier, a major gender-related issue in the use of public transportation is the presence and frequency of sexual harassment cases. According to the Tbilisi urban report,[29] sexual harassment and other acts of violence against women are widespread.

Safety and security are often primary factors influencing women's mobility preferences. The Tbilisi study refers to a 2014 ADB survey[30] in which 45% of female respondents said they had been sexually harassed on Tbilisi's public transportation, particularly in the metro or near the stations. The survey also revealed that 70% of all female respondents said they had been touched, followed, or stared at in a way that made them feel afraid. It is important to emphasize that discrimination and violence in many cases originate from unregulated drivers within the transport system.

A major challenge for vulnerable groups, including internally displaced persons and women from low-income groups, is the unreliability of the intercity transport service. Consultations with stakeholders highlighted the concerns expressed by CSOs regarding the problems faced by these groups due to the informal structure of the service. Particularly in rural areas, members of these groups use buses to get to work, for example in agricultural fields. However, the lack of formalization means that it is unsafe to find a driver willing to transport them back, and there is uncertainty about the costs involved. In many cases, buses are not willing to take them back, especially if there are not enough passengers to make the trip financially viable.

Despite these current problems in terms of safety, security, and reliability, these vulnerable groups are significant users of intercity transport services and use them on a daily basis. This is particularly evident in rural areas, where the household's private vehicle is generally used by men for work-related travel. In contrast, women rely heavily on the bus to commute to work, accompany their children to school, reach health facilities, and access other essential services.

---

29   CEE Bank watch Network. 2022. Urban Public Transport Reform in Tbilisi.
30   ADB. 2014. Georgia: Sustainable Urban Transport Investment Program - Tranche 1 Initial Environmental Examination.

Improving the safety and reliability of public transportation, especially in rural areas, will have a differential impact on the quality of life of these groups.

Conclusions and Recommendations

Change in ticket prices

The design of a suitable intervention to cushion market price dynamics first requires the determination of the targeted beneficiaries or the social welfare function or costs to be optimized. Considering that passenger behavior generally changes with ticket price fluctuations, government authorities should create a set of protection measures and incentives to ensure that no one is left behind.

It is important to tailor the implementation of measures to the specific context of intercity transport buses in Georgia.

(i)   **While in theory middle- and high-income groups might be more sensitive to a fare increase, the reality is that these groups do not currently use the service due to nonprice-related concerns** such as safety, security, efficiency, and reliability. Therefore, government measures for these income groups should prioritize addressing these existing issues and communicating effectively about them, rather than focusing primarily on mitigating a potential fare increase.

(ii)   **Conversely, low-income groups rely heavily on this service[31] (around 556,000 users) as it is their primary and essential means of transportation, despite facing challenges related to security, reliability, and efficiency.** For this demographic, the main concern revolves around a possible fare increase. Although such an increase will not significantly change the use of the service, it could worsen their economic problems and affect purchasing power in other, less essential areas. Government interventions for these groups should focus on addressing affordability issues while alleviating the nonprice-related concerns they currently face. Table 17 provides a summary of the potential impacts of fare increases and proposes appropriate government responses for each group's reaction.

---

[31]   See Figure 28.

Table 17: Possible Impacts of a Ticket Price Increase on the Affected Population Groups

Population Group	Intercity Users	Possible Impact of a Price Increase	Expected Response	Government Possible Intervention
Low- to low-middle income groups	• 37.4% are below GEL1,040 (mean income) • 40.5% are between GEL1,040 and GEL2,080[a]	Negative effect on disposable income, as they have a more inelastic demand: Low-income groups are less likely to change their demand in response to price changes.	Reduce purchasing power in other sectors, as they have limited access to transportation alternatives. The proportion of their total budget on transportation will increase.	Financial assistance, e.g., through cash transfers to be used for transportation services.
Upper middle–income and high–income groups	12% are above twice the median salary (GEL2,080)[a]	Nonsignificant impact on disposable income. In the Georgian context, they usually rely on private cars for commuting. They hardly use public transportation. This is not primarily due to high prices, but to concerns related to service aspects such as safety and reliability. A price increase would not affect them significantly.	In general, they could switch to private transportation alternatives. However, in the Georgian context, they are not expected to change their behavior as a result of price fluctuations as the nonuse of this service is related to other issues.	Showcase the benefits of reform that address their current issue to encourage the use of public transportation: Implement integrated mobility solutions, increase marketing and information campaigns to raise awareness about the improvement in service quality, safety, and transparency.
Students, youth, people with disabilities, elder population	• 21% students • 24% below 18 years old • 16% above 65 years old[a]	Negative impact on disposable income as they have a more inelastic demand.	Reduced purchasing power, as the proportion of their total budget spent on transportation increases when ticket prices rise.	Provide financial assistance, e.g., through cash transfers to be used for transportation services.
Rural population	61% of intercity users[a]	Demand from long-distance travelers is more elastic, i.e., they are likely to change their demand behavior.	Likely to switch to private cars	Showcase the benefits of reform that address their current issue to encourage the use of public transportation: Integrated mobility solutions, marketing and information campaigns to raise awareness about the improvement in service quality, safety, and transparency.
Low and middle-income women	Income by gender is not available	Negative impact on disposable income for those who have access to transportation. Lower-income women have no access to public transport.	Reduced purchasing power as they have limited access to alternative transportation means. The proportion of transportation expenses in their total budget will increase.	Financial assistance, e.g., through cash transfers that can be used for transportation services.

[a] Based on National Statistics Office of Georgia. Geostat Database. https://www.geostat.ge/en (accessed 5 March 2024).
Source: Authors.

Among the various policy instruments that can be used to address these issues, studies on the economics of public transport suggest that price interventions are more efficient if they are based on a careful understanding of market dynamics, especially the demand elasticity of transport services in the country context.[32] For example, monetary interventions to mitigate price increases are more beneficial for the population with elastic demand, as a fare reduction would trigger higher demand for the service.

Box 2 gives some alternative policy interventions that could be considered when designing policy interventions related to affordability. The measures are provided taking into account the government's approach to intervention in the sector as communicated during stakeholder consultations: *to let private entities operate without direct government intervention in the fare.*

Cash transfer programs

Cash transfers provide direct financial support to individuals or households. This can be in the form of consumption vouchers or coupons for transportation services only, preloaded transportation cards, cash transfers linked to expense monitoring procedures, among others. This will also contribute to the traceability of users and better monitoring of demand on the routes. The main objective of this type of instrument is to reduce the cost of the service for the vulnerable population and address income inequalities.

The alternatives mentioned above can mean an improvement in the ticketing system, such as the integration of electronic means of payment for the purchase of tickets, connected apps, to name a few.

Box 2: Country Experiences with Cash Transfers

Niger's solution to a poorly developed financial system and the high cost of distributing physical money was to transfer money via mobile phones. This proved to be cheaper and time-saving for users. Beneficiaries of the program have better access to a more diverse diet and greater bargaining power for women.

In the United States (US), the Transportation Voucher Program was implemented in five counties in Texas. It consisted of a voucher targeting people aged 60 and over to improve their access to public transportation.

Also in the US, the Transportation Wallet for Residents of Affordable Housing Program provided financial assistance to low-income participants in the form of a prepaid visa card that could be used for public transportation.

Sources: GSMA. Women's Trust of Mobile Money and Agents in Nigeria and Senegal. https://www.gsma.com/solutions-and-impact/connectivity-for-good/mobile-for-development/wp-content/uploads/2024/01/Womens-Trust-of-Mobile-Money-and-Agents-in-Nigeria-and-Senegal_Eng.pdf; and Deep East Texas Council of Governments (DETCOG). 2019. *Transportation Voucher Program Final Report.*

---

[32]   J. S. Dodgson and N. Topham. 1987. Benefit-Cost Rules for Urban Transit Subsidies: An integration of Allocational, Distributional and Public Finance Issues. *Journal of Transport Economics and Policy.* 21 (1). pp. 57–71.

For this reason, cash transfer programs in developing countries with limited financial infrastructure often require the physical distribution of cash in small denominations in remote rural areas, which can lead to higher costs for the implementing agency. If this is the case, some auditing mechanisms could be introduced where the recipient has to prove that they are spending the money as expected.

In Georgia, a voucher policy was introduced in Tbilisi in 2011.[33] This could serve as a reference for the development of the new scheme.

Further research needs to be conducted to develop a suitable tool that fits the policy objectives and the new Georgian intercity transport landscape.

Subsidy programs

Price interventions in the form of subsidies are commonly used to address affordability issues by lowering the cost of services. Subsidies can be channeled directly to beneficiaries, known as demand-side subsidies. When these subsidies follow an effective eligibility criterion to screen low-income households, they have a similar effect to the cash transfers mentioned above. The main difference is that the subsidies are price interventions and not direct cash monetary transfers.[34]

In developing countries, such as Georgia, the provision of public transport services is usually in the hands of private companies or individuals. Since the government's approach is not to intervene in the fares set by the transport operators, the focus of this recommendation depends on the price control mechanisms that the government could put in place. Again, the aim of a subsidy would be to make intercity transport more affordable, which would have an impact on low-income and infrequent users.

The effect of subsidies varies depending on the policy tool used (footnote 34). Box 3 summarizes countries' experiences within the implementation of different types of subsidies.

Box 3: Country Experiences with the Implementation of Subsidies

Financial support in the form of subsidies can be implemented through different instruments. Some examples from the countries' experiences are as follows:

Means-testing transfer financed by general taxation: the case of Bulgaria, Chile, Mexico, Spain, and other countries where special categories of passengers travel for free or at a lower price than the standard fare.

Flat fare structure: uniform price regardless of the distance traveled.

Source: N. Estupinan et al. 2007. Affordability and Subsidies in Public Urban Transport: What Do We Mean, What Can Be Done?. *Policy Research Working Paper Series*. World Bank.

---

[33] Georgian News Agency. 2011. Tbilisi Government Issues Transport Vouchers.

[34] N. Estupinan et al. 2007. Affordability and Subsidies in Public Urban Transport: What Do We Mean, What Can Be Done?. *Policy Research Working Paper Series*. World Bank.

Another important aspect that needs to be carefully analyzed is the source of subsidy funding. This should be planned and aligned to the financial capacity of the government.

In any case where a policy instrument is used (subsidy or direct cash transfer), it is important to **conduct a preliminary study** to determine the target groups, their socioeconomic characteristics, and the way in which they use the service; and (ii) **implement continuous monitoring and fiscal control** to ensure that the financial aid reaches the target group and that the expected distribution results are achieved.

Gender-specific recommendations in relation to changes in ticket prices

As mentioned earlier, the impact on affordability of services will have a relevant gender effect as women rely heavily on the intercity network compared to men. To mitigate the disparities of fare increases, government initiatives should include a gender-sensitive approach.

The system should meet the mobility needs of women by providing financial assistance to reduce the burden of transportation expenses for the entire family.[35]

Targeted assistance programs specifically designed to support women commuters from low-income households should be considered to ensure that they are not overburdened by fare increases. Transfers can take the form of subsidies, vouchers, or direct financial assistance.

Further consideration could also be given to women living in remote rural areas and women who are responsible for childcare and may be in a more precarious economic situation.

Use of services

The following measures are designed to further promote the use of intercity bus services (especially for middle- and high-income groups) by making them more attractive to users:

(i) **Monitor and ensure safety and security.** This is a major concern across all income groups. For middle- and high-income groups, it is a significant deterrent to using the intercity bus service. At the same time, it remains one of the foremost challenges for low-income groups who lack alternative transportation options. Measures to improve safety and security should specifically address issues such as sexual harassment, road safety, and other relevant factors. Implementation of these measures will bring significant benefits to current users and potentially attract new users from other segments of the economy. Government action should focus on the following, among other things.

- Ensure that drivers comply with road safety measures.
- Install and maintain surveillance systems on buses and at stations and investigate cases of harassment or violence. This would act as a deterrent and provide evidence when incidents occur.

---

[35] A. Iluridze and T. Verulashvili. 2021. *Women and Mobility: Gender Aspects of the Women Daily Movement.* Public Defender (Ombudsman) of Georgia and GIZ.

- Conduct public awareness campaigns for passengers and drivers on acceptable behavior, reporting mechanisms, and the consequences of harassment.
- Establish an emergency hotline that is secure and easily accessible.
- Perform regular audits.
- Protect whistleblowers who report safety concerns or incidents to ensure that those affected feel safe to come forward with information.

(ii)   **Monitor and ensure reliability, efficiency, and transparency.** This is another major concern for both users and nonusers of public transport. The government should ensure that the measures planned as part of the reform effectively address the current problems. This must be supported by monitoring, reporting, and continuous improvement.

(iii) **Improve inspection protocols,** guaranteeing that the new safety measures and behavioral norms on board buses are being respected. Allocate resources to conduct regular and thorough inspections of vehicles and operating practices, promoting a culture of compliance with regulations.

(iv) **Monitor and evaluate the level of compliance.** Develop a framework to systematically track the pace of reform implementation and ensure that service providers are adhering to the new regulations. Define performance indicators for compliance, safety, environmental impact, and overall operational efficiency.

(v)   **Carrying out marketing and information campaigns.** To raise public awareness of the improvements that have been introduced into the system following the reform, these campaigns should emphasize the benefits of using the intercity public transport, such as improved safety, environmental benefits, better route connectivity, and cost savings compared to private vehicles. Such campaigns will help to overcome negative stereotypes or misconceptions that may currently exist.

(vi) **Create integrated mobility solutions.** Consider developments that integrate intercity and urban transportation systems as well as different modes of transportation such as buses, bicycles, trains, and connections to airports. These measures make it easier for passengers to connect from their origin to their destination. Some examples of developing countries with integrated systems are Colombia in Bogotá,[36] Brazil in Curitiba,[37] Mexico in Mexico City,[38] and Nigeria in Lagos.[39] Intercity bus service is more likely to be successful in markets where rail and air services are limited or expensive and where there is a large population base and a high degree of intercity travel.[40]

[36] L. A. Guzman, D. Oviedo, and R. Cardona. 2018. Accessibility Changes: Analysis of the Integrated Public Transport System of Bogotá. *Sustainability*. 10 (11). p. 3958. https://doi.org/10.3390/su10113958.

[37] Curitiba, Brasil. n.d. BRT Case Study.

[38] Gobierno de la Ciudad de Mexico. The Mexico City Integrated Transit Card.

[39] R. Musbau. 2021. Lagos and the Development of Integrated Transport System. *The Guardian*.

[40] J. E. Purdy and L. A. Hoel. 1980. An Evaluation of the Effect of Level of Service and Cost on Demand for Intercity Bus Travel: Final Report. United States Department of Transportation, National Transportation Library.

(vii) **Develop customized commuting solutions.** Introduce shuttle services that connect residential areas with major transit hubs, providing a convenient last-mile solution.

(viii) **Define and communicate formal timetables.** Ensure that users can rely on public transport if they want to travel on time.

Gender-specific recommendations regarding changes in the use of services

In Georgia, the Tbilisi Transport Company has introduced initiatives focusing on equal opportunities for women in the sector as well as a safety policy. In 2022, the Gender Advisory Services was introduced with the support of the European Bank for Reconstruction and Development.

For intercity transport, an expansion of cooperation between the Transport Agency and the Gender Advisory Services initiative would positively influence policy design and government action to promote gender inclusiveness in the transport sector.

In addition to the above measures, the following recommendations could be implemented to create a safer environment that minimizes gender discrimination:

(i) **Increase the labor participation of women in transportation services.** This will reduce hate crime and violence while increasing gender participation in the sector.

(ii) **Strengthen the typification of discriminatory behavior and enforcement of policies.** Define sanctions for people who commit gender-based violence. Provide enforcement mechanisms to ensure that policies are implemented.

(iii) **Increase police presence and checks, e.g., through cameras, on buses, and at bus stations** to ensure that behavioral measures are adhered to and to prevent deviations.

(c)   Variable 2: Impact on Compliance Costs for Service Providers

Summary of Outputs and Findings

Registration requirements, safety measures, technological standards, and other service improvements introduced by the reform would raise compliance costs for all operators, as they would have to comply with the new regulations in order to continue operating in the market.

According to the Regulatory Impact Assessment (RIA) conducted by GIZ, there are 4,125 people providing passenger transportation services on intercity routes across Georgia, 94% (3,903 units) of them self-employed drivers and 222 of them legal entities. Regarding the service characteristics, accurate and complete data on timetables, passenger flows, and intercity routes were not available. Therefore, the assessment was built on the basis of stakeholder consultations with the service providers presented in the RIA.

On average, 42,700 people travel daily on intercity routes operated by licensed (formal) transport operators, and the average number of trips is 3,879 (Table 18). However, the estimates are not sufficient to determine the actual number of long-distance trips and passengers carried. According to the consultations, the total share of passengers transported by *kishniks* (unregistered drivers) is between 30% and 80%, depending on the route and location.[41]

Table 18: Data on Transportation Service Providers

Total Registered Service Providers	4,125
Registered self-employed drivers	**3,903**
Owning 1 vehicle	3,478
Owning 2 vehicle	295
Owning 3 vehicle	76
Owning 4 or more vehicles	54
Registered company drivers	222
Passengers Traveling with Registered Drivers (Daily)	**42,700**
Average Trips/Routes per Day	**3,879**

Source: GIZ. Regulatory Impact Assessment. https://pbo.parliament.ge/estimates/regulatory-impact-assessment.htm.

Assessment of Compliance Costs

To assess the financial implications of the reform on service providers, two compliance scenarios were defined. The *baseline compliance* scenario considers the case of operators with an annual salary of less than GEL30,000 (exempt from income tax) who do not need to invest in a new vehicle. The *upgrade compliance* scenario considers the case of "small entrepreneurs," who are taxed at 1%, and the acquisition of a bus that meets the Euro 5 standard. Overall, the bus acquisition implies the highest upfront cost that service providers would incur to comply with the new regulations.

Two income levels were considered: a lower limit defined by an annual salary of GEL30,000 and an upper limit up to an annual salary of GEL500,000. This can be observed in Table 19.

Table 19: Comparison of Two Baseline Compliance Scenarios

Quantification Scenarios	Scenario 1: Baseline Compliance Total costs S1 = Permit cost + Insurance + Estimated transport manager cost
	Scenario 2: Upgrade Compliance Total costs S2 = Total costs S1 + Bus acquisition + (Income) Tax

S = scenario.
Source: Author's own elaboration based on Regulatory Impact Assessment data.

---

[41] Deutsche Gesellschaft für Internationale Zusammenarbeit (GIZ). 2021. *Regulatory Impact Assessment.*

The cost estimates are based on available data and assumptions (Table 20).

Table 20: Inputs and Assumptions to Estimate Compliance Costs for Service Providers

Assumptions ($)		Comments
Transport manager annual salary	7,940	*Source*: RIA estimates. A transportation manager can only manage four operators at a time, with a total number of vehicles owned not exceeding 50. *Assumption*: A self-employed driver is assumed to pay one-fourth of the estimated annual salary.
Income tax rate	1%	*Source*: Georgian Tax Code. A 1% income tax is applied to self-employed drivers who enjoy the status of "small entrepreneur." *Assumption*: The calculations were made for a lower limit of GEL30,000 and an upper limit of GEL500,000.
Cost of acquiring a new vehicle	39,070	*Source*: Market price (myauto.ge) *Assumption*: The requirements for Euro 5 vehicles will apply.
Insurance costs	100	*Source*: Estimate based on local consultations.
Proof of financial status	**3,320** for the first vehicle **1,845** for each additional vehicle	As part of the provisions of the European Union regulation, the reform requires proof of the financial capacity of transport operators. However, several other alternatives are also being considered as a means of financial resources. (i) Any assets (including monetary values) and/or annual reports and/or bank guarantees. (ii) Vehicles owned by the permit holder and/or operated under the binding legal form of leasing are also considered as financial resources and the obligation to have additional resources does not apply to these vehicles. (iii) Real estate owned by the permit holder can also be considered as financial resources.[a]

RIA = Regulatory Impact Assessment.
[a] Government of Georgia. Explanatory note on the draft law of Georgia. Amendments to the Law of Georgia "On Motor Transport."
Source: Author's own elaboration based on RIA data.

It is worth noting that the financial proof requirements for drivers were not included in the total upfront cost calculations. This is due to the strategic alternatives introduced by the government that could make it easier for service providers to provide the required proof of solvency. The implementation of these alternatives has streamlined the process and makes it easier for providers to meet the necessary financial obligations without adding to the initial financial burden on drivers.

Table 21 provides estimates for the compliance scenario. It considers the implementation period until 2025, during which the cost of the permit will first rise to GEL100, and then to GEL500. Under the new regulation, one transport manager can be shared by four service providers.

Table 21: Cost Estimates for the Baseline Compliance Scenario

Scenario 1	Baseline Compliance Costs ($)	
One time cost	**Cost until 2025**	**From 2025 onward**
Permit fee to operate	37	185
Annual fixed costs	**Annually** ($)	
Vehicle insurance costs	100	
Hire a transport manager	1,985	
Individual drivers pay 1/4 of the annual salary		
Scenario 1 Total Cost	**2,122**	**2,269**

Source: Author's own elaboration based on Regulatory Impact Assessment data.

The upgrade for the compliance scenario includes the one-time costs for acquiring a new vehicle and the periodical income tax payments, as seen in Table 22.

Table 22: Cost Estimates for the Upgrade Compliance Scenario

Scenario 2	Upgrade for Compliance ($)	
One time cost		
Cost of acquiring a new vehicle in compliance with Euro 5	39,072	
Annual costs		
Income tax, if income = GEL30,000	111	
Income tax, if income = GEL500,000	1,845	
Scenario 2 Total Cost	**Including taxes**	**Including taxes + vehicle acquisition**
Costs with income tax, if income = GEL30,000	**2,232**	**41,304**
Costs with income tax, if income = GEL500,000	**3,967**	**43,039**

Source: Author's own elaboration based on Regulatory Impact Assessment data.

Bus station fee

Service providers must pay a fee to bus station owners for carrying passengers to their station. The amount of the fee is currently variable and depends on the revenue from the ticket sales and is usually determined on the basis of an agreement between the parties.

Currently, the bus station fee varies between 1% and 20% of the ticket revenue and depends on the distance and demand on a particular route. Normally, drivers have to pay a higher fee for the most popular routes (footnote 41).

The reform aims to put an end to this arbitrary fee setting and requires that the same fee applies to all bus operators within that station. The fee will be set by individual bus stations, but there will be a cap of 20% of the fare. This will help to create a much more competitive environment among service providers while limiting the discretion of bus station owners in setting this fare.

For the purposes of this study, this cost was not included because it is not an upfront investment but rather a variable cost, depending on the level of operation. It does not increase the operators' reform-related financial obligations, as they are paid from ticket revenues.

Number of Affected Workers

The escalating compliance costs and the barriers identified in the previous section could jeopardize the viability of several operators and expose them to the risk of losing their business if they are unable to adapt to the requirements of the reform. This aspect is particularly important for unregistered drivers, who make up a substantial proportion of total service providers. It is worth noting that their service will also depend on the design of new routes and the allocation of authorized stations within their jurisdiction.

The estimated total number of unregulated drivers potentially affected by the reform adds up to 4,000 workers. This estimate is based on considering inputs from local consultations, according to which unregulated drivers make up 40% of the total workforce, while regulated workers total 6,000. This can be seen in Figure 38.

Figure 38: Georgian Workers for the Intercity Transport Labor Market, 2021

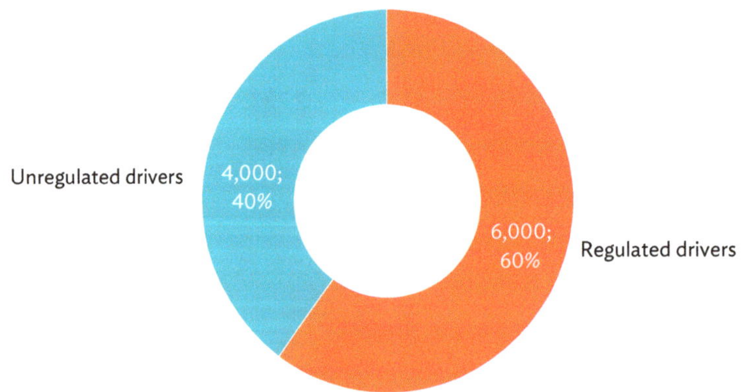

Source: Local consultation estimates.

To put it in perspective, this translates to 5% of the total workforce in Georgia's transportation and storage sector, which in turn constitutes 6% of the country's total workforce, as seen in Figure 39.

Figure 39: Participation of Transport and Storage Workers in the Total Georgian Workforce, 2021

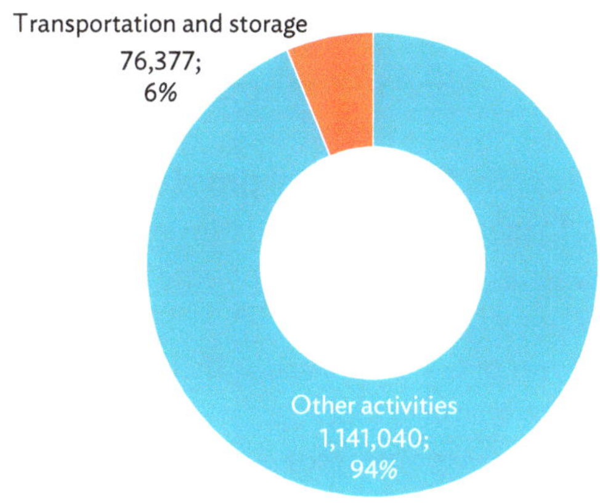

Source: Geostat. 2021. Labour Force Survey. Employment by Branch.

Informal workers face not only financial barriers to formalizing their activities, but also administrative challenges. Considerations regarding the transition window to minimize their costs and ensure their continuity in the market will play an important role.

Conclusion and Recommendations

From the above analysis, it is clear that the biggest financial challenge for service providers is the purchase of Euro 5-compliant vehicles. Within the implementation window being considered by regulators, changing their fleet implies a considerable upfront investment, especially for self-employed drivers. In addition, operational requirements—such as safety regulations and licenses and administrative procedures such as fiscal obligations—will challenge their capacity to obtain the certifications and adapt their activity to the new regulations.

Aside from financial considerations, drivers may be reluctant to comply with intercity transport reforms because they have a number of non-cost-related concerns. For example, there may be apprehension stemming from years of operating as unregulated drivers and fear of potential repercussions. In addition, a lack of clear understanding of the reform's requirements could pose an additional challenge for drivers. Addressing these non-cost-related concerns is crucial for the successful implementation of intercity transportation reforms so that drivers feel confident and well-informed and are more likely to adopt the new regulatory framework.

The following recommendations aim to mitigate the impact of compliance costs on drivers and create a favorable environment for compliance with regulatory standards in the context of intercity reform in Georgia.

(i)   **Loans for new vehicle acquisition.** Preferential rates, flexible guarantee requirements, and other financial support will be key to enable drivers to access new vehicles in compliance with the reform. This can be provided by the government or as a blended measure in collaboration with international organizations and the private sector.

(ii)   **Targeted assistance program for financial proof,** to address the associated requirements for financial guarantees for service providers. Through this financial support, the government is ensuring a smooth transition, preventing economic barriers from excluding individuals who are less solvent and may face additional financial challenges to remain in the market. This targeted assistance not only helps those with limited financial resources, but also demonstrates a commitment to creating a fair and accessible regulatory environment for all participants in the intercity transport sector.

(iii)   **Information and resource centers.** Create centralized information and resource centers that inform drivers about available support mechanisms, compliance procedures, and cost-saving strategies. Improve communication channels to ensure drivers are well-informed about the procedures and paperwork they need to complete. This is even more so for adults with low levels of education and the self-employed in the informal business, as they may find it difficult to overcome the administrative hurdles to align their activity to the new regulations. Clear communication can also help informal workers to cope better with the changes.

(iv)   **Training subsidies.** Targeted financial support that ensures drivers have access to affordable, high-quality training to meet evolving safety and regulatory requirements. These training programs can also improve the skills of informal workers and ensure they are well-equipped to meet compliance standards.

The estimated number of workers affected is considerable but appears to be manageable if appropriate measures are taken. Ensuring the effectiveness of carefully tailored strategies can turn this challenge into an opportunity to promote a smoother and more inclusive transition for all involved parties. To minimize the financial burden and operational challenges and reduce the potential resistance that the new formal scheme will trigger, support and assistance for these workers during the transition is key to ensure an efficient and smooth process.

Below are specific recommendations to minimize the impact on the most vulnerable informal workers. To make interventions more efficient, it is important to better understand the profile and characteristics of the target group. Older workers, workers with large families, workers with low levels of education and workers who have few opportunities to engage in alternative economic activity are likely to need support through tailored policy interventions.

(i)   **Transitional support programs.** Introduce programs to help informal bus drivers meet the new compliance standards. This could include financial assistance, training initiatives, and resources to ease the transition period.

(ii) **Flexible compliance timelines.** Offer a phased approach to compliance deadlines to give informal workers more time to adjust to the new requirements. This flexibility can mitigate immediate financial burdens and promote a smoother transition.

(iii) **Legal assistance and counseling services.** Set up legal assistance and counseling services to inform informal workers about their rights, responsibilities, and opportunities to formally enter the market. This can help remove uncertainty and ensure fair treatment during the transition.

Gender-specific recommendations

The intercity transport reform is an opportunity to redefine the landscape of service providers and make it more inclusive for women and minorities.

Policies should encourage the participation of women among drivers and other workers in the sector.

The recommendations aim to strike a balance between regulatory compliance and supporting the livelihoods of informal workers in the face of increased costs caused by the reform. They also aim to improve the new formal landscape of intercity service providers, making it more inclusive for all genders.

Through targeted and collaborative interventions, the government can facilitate a smoother transition to an improved intercity service, while maintaining the objectives of the regulatory changes.

(d)   Variable 3: Impact on Local Communities

Summary of Model Outputs

The reform will also have an impact on the current landscape of bus stations, which may affect the local communities that rely on them. Two main impacts have been identified: (i) impact on informal workers who depend on bus station operations (both formal and informal) and (ii) impact on remote villages that rely on informal bus stations to access bus transport services.

(i)   **Impact on informal workers in the vicinity of bus stations**

The reform could affect informal workers operating in and around bus stations. Whether it is the closure of existing stations, the inauguration of new stations, or the upgrading of existing station classifications (first, second, and third class), these changes will reshape the commercial landscape of the station.

Due to the introduction of this reform, it is likely that commercial activity around the stations will be driven by formal businesses such as stores, cafeterias, and supermarkets. This will leave little room for informal vendors who may currently still rely on the flow of passengers around stations. As a result, informal vendors may be displaced and have to find another place where they can continue to sell their products informally, or they may be absorbed by new formal job opportunities.

The impact of these changes depends on the geographic location of the bus stations. In large cities, informal workers are likely to move to similar activities in other locations. In contrast, changes in rural and remote areas could have a greater impact on surrounding communities due to limited alternative options.

This assessment aims to identify the number of informal vendors potentially affected by the reform. Although there is limited public data on the informal sector, some references were found in the RIA developed by GIZ. According to the RIA, "there are few first-class stations. First class facilities are the most advanced, including space for an information board, administrative purpose rooms, medical point, sanitary and hygienic block, rest rooms for drivers, among other services. However, currently most of the bus stations are third class (102), meaning that they only have a platform for busses to stop, a ticket office and a minimal waiting area" (footnote 41). It can be assumed that higher-class stations offer more space for formal services and thus also for formal jobs.

Estimates of affected informal workers

To estimate the number of potentially affected informal workers in the station area, assumptions about their current work dynamics were combined with information from the RIA (Table 23).

Table 23: Assumptions for the Calculation of Potentially Affected Informal Workers in the Vicinity of the Station

Assumptions	Amount
Informal vendors in the vicinity of stations	10 workers
Activities developed by micro, small and medium-sized enterprises	Sale of food and beverages, other local products, and souvenirs
Declared bus stations (data from the Regulatory Impact Assessment)	160

Source: Author's own elaboration based on information from Regulatory Impact Assessment.

Assuming that 10 informal vendors work near a bus station, 1,600 informal workers could be affected by changes in bus station characteristics and geographic distribution.

However, there are several unreported stations that could also attract informal activity in their vicinity.

The number of workers affected will ultimately depend on each municipality's strategy for improving the efficiency of the new routes and the permitted number and classes of bus stations in their jurisdictions.

(ii)   Impact on remote villages

Currently, many remote rural villages rely on the existence of informal bus stations to facilitate their intercity bus transport mobility. The closure of these stations could have far-reaching consequences for these communities, especially considering that there are few alternatives in such areas. As a result, residents may find themselves forced to travel additional distances, likely on foot, to reach the nearest formal bus station. This underscores the crucial role that informal bus stations play in the accessibility of these remote villages. Figure 40 shows the average household expenses in intercity transport.

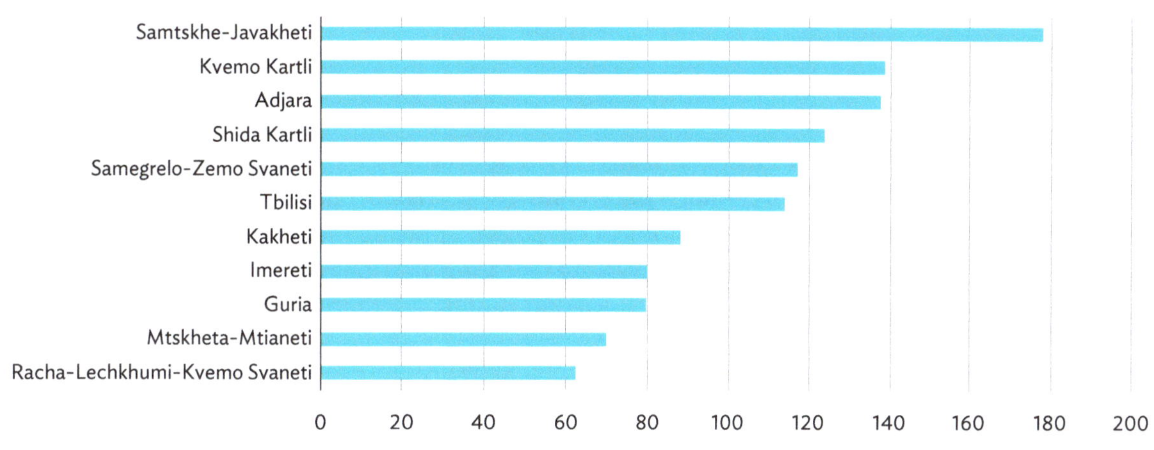

Figure 40: Average Household Expenses in Intercity Transport

Sources: National Statistics Office of Georgia. Geostat Database. https://www.geostat.ge/en (accessed 5 March 2024).

According to the RIA, the 160 bus stations are mainly concentrated in Imereti (32), Kvemo Kartli (28), and Tbilisi (24). **It is interesting to compare these figures with the top three regions where households report spending on average a higher amount of their budget on intercity transport services** (data collected by Geostat).[42] This can be observed in Figure 41. These are Samtskhe-Javakheti (with only four registered stations), Kvemo Kartli, and Adjara. It is reasonable to assume that the regions that spend more on the service concentrate more frequent users, offer better station facilities, and have more organic opportunities to develop economic activity in their area.

---

[42]   National Statistics Office of Georgia. Geostat Database. https://www.geostat.ge/en.

Figure 41: Regional Distribution of Bus Stations According to Classes

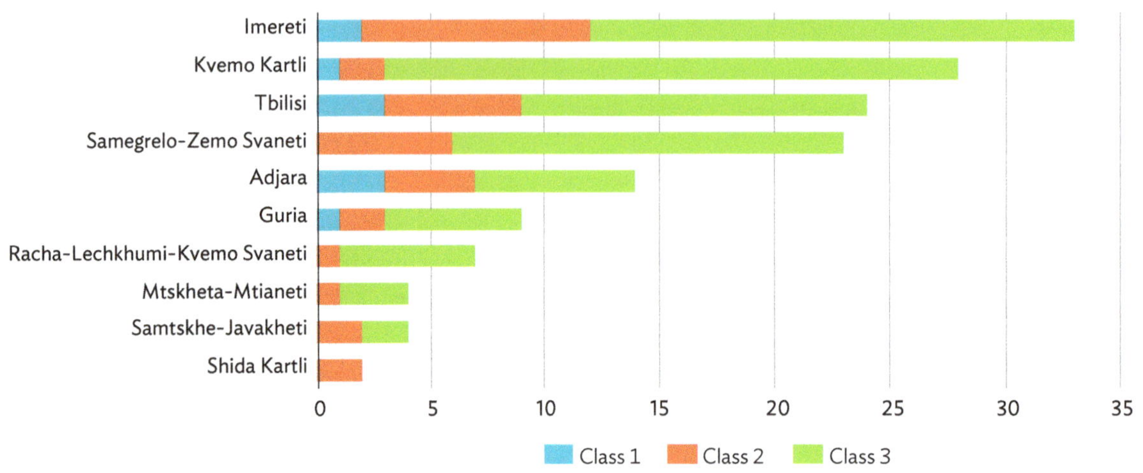

Source: Regulatory Impact Assessment.

Gender aspects and the assessment of vulnerable groups' issues

It is important to note that the informal workers most affected by these changes are from the lowest-income brackets, which has a particular repercussion on women. These people are already struggling with substantial economic challenges, implying that any disruption to their economic situation could have serious consequences for their well-being. It is therefore important to take mitigation and preventive measures to ensure that they are not left behind in this transition.

Furthermore, the closure of informal bus stations will most likely affect women, young children, and the older people in remote rural villages, who are the most frequent users of this service.

Conclusion and Recommendations

The local communities will be indirectly affected by the new bus station landscape created by the implementation of the reform. While positive socioeconomic impacts can be expected for residents of remote and economically disadvantaged areas, such as increased employment opportunities and the potential for local economic growth through the enhanced influx of travelers, there is also a risk for informal commercial activities. Informal workers such as local artisans and food vendors who depend on the current dynamics of stations could face challenges due to infrastructure changes and relocation of stations. There is also a risk for residents of remote rural areas that they will be disconnected if they do not have access to a nearby bus station.

Given the significance of the project, the impact on the estimated 1,600 informal workers affected will depend on the government's policy management. The role of local authorities and municipalities will be critical in effectively addressing potential negative impacts on the jobs of local informal vendors and maximizing the advantages for the population living in remote areas. Some recommendations for policy interventions are as follows:

(i)　**Plan for bus stations and alternative modes of transportation.** Ensure that formal bus stations are strategically located at a reasonable distance from nearby villages. For villages farther away, establish alternative transportation to facilitate access to the bus station.

(ii)　**Provide financial incentives to operators to ensure the provision of services on less profitable routes.** This is to compensate them for operating less profitable routes and incentivize them to plan routes that avoid overcrowding on more popular routes.

(iii)　**A dedicated space for local artisans and workers.** Require bus stations to set up a dedicated area for local artisans and workers, with a certain quota allocated to women. This initiative will ensure that these individuals have a designated and accessible space to showcase their products and make a living amid the changes associated with the reform. Box 4 highlights an example of this in highway service stations in Georgia.

Box 4: The Example of Highway Service Stations in Georgia

The case of highway service stations in Georgia offers an important parallel to the effects on informal workers discussed in this context. When old roads where informal workers could sell their goods at intersections or stop lights are replaced by highways, these employment opportunities disappear. In response, the Georgian government has designated special areas at highway service stations for local artisans to continue selling their goods.

While this initiative serves as a potential model for mitigating the impact on informal workers at bus stations, it is crucial to acknowledge that its success was limited. A major challenge was that it was impractical for workers to travel to the service station daily to sell their goods.

Learning from this, future measures should be carefully considered to take such circumstances into account and ensure their effectiveness.

Source: Deutsche Gesellschaft für Internationale Zusammenarbeit (GIZ). 2021. Regulatory Impact Assessment.

(iv)　**Promote the employment of local workers.** Require bus stations to give priority to employing local workers in the shops and businesses within the station.

(v)　**Community development initiatives.** Implement community development projects that focus on skills building, education, and improving local infrastructure. This can provide community members with new skills and improve the overall quality of life in the vicinity of bus stations.

(vi)　**Cultural and recreational spaces.** The area surrounding the stations could also be enhanced with cultural and recreational spaces to create community hubs and attract visitors. This could potentially lead to new economic opportunities.

These strategies aim not only to mitigate the negative impacts, but also to foster a resilient community, promote a just transition where new employment opportunities are created, and improve the overall well-being of the neighborhoods potentially affected by the new regulations.

(e) Recommendations on Activity 3

Table 24 summarizes the key recommendations for each measure based on the identified impacts and the corresponding significance assessment.

The details of the significance assessment can be found in Appendix 2.

Table 24: Summary of Findings, Recommendations, and Impact Significance Assessment

Variable	Type	Significance	Impact Summary	Recommended Measure
Variable 1: Distributional impacts on service users	RISK/ OPPORTUNITY	MODERATE	Low-income groups: 15% – 556,000 users negatively affected in their disposable income. Positive impact on service use if issues are properly addressed. Middle- to high-income groups: no impact on disposable income. Potential increase in demand if nonprice-related issues are properly addressed.	• For low-income groups: financial assistance through cash transfer programs. Take care of safety and reliability issues. • For upper-middle income and high-income groups: address and communicate about current safety, reliability, quality, and efficiency issues.
Variable 2: Impact on compliance costs for service providers	RISK	MINOR	4,000 drivers at risk mainly due to the cost of acquiring a new vehicle ($39,000)	• Preferential financing and de-risking programs to facilitate access to loans for new vehicle acquisitions. • Targeted assistance to provide financial proof. • Information, resource and training centers at the regional level.
Variable 3: Impact on local communities	RISK	MINOR	1,600 informal workers working around the bus station are potentially at risk. Remote villages are potentially at risk of being disconnected.	• A dedicated space for local artisans and workers at bus stations. • Promotion of local employment at bus stations. • Planning and mapping alternative transportation measures for remote villages to access bus stations. • Financial incentives for operators to use routes in low demand.

Source: Authors' elaboration.

The intercity transport reform has the potential to bring both negative challenges and costs for affected stakeholders and opportunities that positively impact the user experience and working conditions and benefit different population groups.

The assessment found that the main impact will be the distributional effects on service users. However, the impact will be quite different depending on the population segment:

(i) **For low- and middle-income groups, the main risk would be the potential increase in fare.** This is because these population groups rely on these modes of transport for essential activities and would then experience a reduction in their disposable income for other, less essential purchases. However, the impact can be mitigated by implementing cash transfer programs following the example of other countries. In addition, these groups could benefit from other improvements of the reform, e.g., in terms of safety and reliability.

(ii) **The upper-middle and high-income groups would not be affected by a price increase as currently do not rely on the service. However, demand from these segments could be encouraged** by leveraging the service improvements through communication campaigns and the introduction of well-managed enforcement controls.

For service providers, the financial costs and administrative challenges are considered to be minor, although it is important to address them effectively. The main upfront costs incurred by service providers are related to the purchase of new vehicles. Cost estimates range from $2,122 for small enterprises that do not need to purchase a new vehicle to $41,304 if these companies need to renew their bus fleet. To support the most vulnerable service providers, targeted financial measures such as preferential financing and de-risking programs can be implemented. In addition, nonfinancial challenges such as administrative procedures to comply with operating licenses can also hinder their continuity. Providing training in these areas as well as information and support centers will help informal drivers (especially the self-employed) to register and stay in the market.

In addition, the new regulated system would offer the opportunity to increase the participation of women in the labor market. This could also have a positive impact on women's safety and the number of harassments in and around public transportation.

Indirect impacts of the reform will also affect the economic activity of workers who rely on informal jobs due to the current dynamics at bus stations. Local authorities should provide support measures to integrate informal workers who could be displaced if the stations are closed or if the reform leads to the establishment of formal stores that leave little room for informal vendors. Policies could mitigate these impacts on people who rely on informal activities by establishing a dedicated space for local artisans; promoting local jobs; creating local, cultural, and recreational spaces; and supporting other community development initiatives.

Finally, it is important to ensure that remote villages are not disconnected from intercity bus transportation. This can be done through planning and mapping as well as alternative transportation measures for remote villages to gain access to bus stations.

Gender issues and opportunities

The reform of the public transport system provides a good opportunity to address several gender-related issues in the transport sector. Women often bear a disproportionate economic burden related to transportation costs resulting from limited transportation alternatives and frequent use, especially when accompanying children and the older people. To alleviate this financial burden, measures should be taken to ensure women have financial access to intercity transportation by increasing support for their mobility-related expenses.

In addition, women are subject to sexual harassment and unreliable schedules, creating an environment that is both unsafe and inconvenient for them. The reform has the potential to address these issues by introducing comprehensive safety measures, ensuring punctuality and reliability, and promoting a culture of respect and inclusivity within the transportation system.

To ensure the successful implementation of these measures and their positive impact on women's experiences in transport, continuous monitoring is essential. This monitoring should focus on assessing the effectiveness of safety measures, addressing recurring issues with the reliability of timetables and ensuring that all stakeholders, including transport operators, drivers, and passengers, are fully meeting expectations.

By addressing these gender-specific issues, the reformed public transportation system will not only provide a safer, more efficient, and reliable mobility solution for women, but will also contribute to a more just and equitable society for all Georgians.

Conclusion

While the assessment has indeed revealed potential negative socioeconomic impacts that could hinder the implementation of climate action in Georgia's energy and transport sectors, it has also uncovered significant opportunities. Successfully mitigating negative impacts and capitalizing on these opportunities is crucial for a just transition.

The study also provides a list of recommended actions that could be implemented by national authorities based on the identified impacts and opportunities and the significance of each of them. Implementing these recommendations will require considerable planning and strategic efforts, such as addressing economic disruption, raising public awareness, and adopting a comprehensive structural approach. The insights provided by this report will guide informed decision-making but implementing these plans can be a complex and significant undertaking.

It is also important to recognize that the findings of this study provide a picture at the sectoral level. For more detailed and nuanced insights, a project-level social and environmental assessment could be useful. This complementary approach would provide granular information that would enable a deeper understanding of the specific impacts and opportunities within individual projects.

Appendixes

Appendix 1. Technical Cards of Socioeconomic Variables

(a) Construction of Renewable Energy Power Plant

Table A1.1: Distributional Impact on Electricity Expenditure in Household Groups

Description	This socioeconomic variable reflects how much the change in electricity prices affects the proportion of household income allocated to electricity expenses. This can help to understand whether the energy transition has the potential to **exacerbate energy-related poverty for vulnerable groups** and provides insight into the potential **economic burden of electricity expenses on low-income households due to price changes.** It considers that **the impact may not be the same for all households and could disproportionately affect certain segments of the population** (distributional impacts). Note: A price increase may lead to a decrease in consumption which may mitigate the actual impact on electricity expenditure. However, this variable aims to assess the impact on current expenditure.
Quantification formula	(1) Change in electricity price: $\bullet$ ΔPrice = New Electricity Price - Initial Electricity Price (2) Change in energy expenditure for each income group "i": $\bullet$ ΔExpenditure_i = ΔPrice * Initial Electricity Consumption_i (3) Initial household income share spent on electricity for each income group "i" (InitialShare_i): $\bullet$ InitialShare_i = (ElectricityExpenditure_i/Initial Household Income_i) * 100 (4) New household income share spent on electricity as a percentage of household income for each income group "i $\bullet$ NewShare_i = (ΔElectricityExpenditure_i/Initial Household Income_i) * 100 (5) Change in share of electricity expenditure for each income group "i": $\bullet$ ΔShare_i = NewShare_i/InitialShare_i
Data sources	$\bullet$ Household income groups and average income of each group per household: National household survey $\bullet$ Average current energy consumption of each income group per household: National household survey $\bullet$ Current electricity price for each income group: EnergoPro Energy
Assumptions	According to the international literature, it is possible that the addition of 1 gigawatt of renewable energy could lead to a slight increase in energy tariffs in the short term due to the initial investment costs. In the long term, however, this could potentially lead to a reduction in energy tariffs as the running costs are lower and the initial investment is recovered. A hypothetical price increase scenario of 5% was used as a basis. Please note that these are very rough estimates and actual figures may vary greatly depending on specific local conditions, government policy, type of renewable energy source, and other factors. To obtain accurate figures, it is best to get primary data from the government.

Source: Authors.

Table A1.2: Job Creation for Renewable Energy Construction Projects

Description	Number of direct jobs created during the construction of the renewable energy (RE) power plants. Although these jobs are temporary, they give workers the opportunity to gain experience and find another job in the future.
Quantification formula	Total jobs = (EC[s] * IC[s])+ (EC[w] * IC[w])+ (EC[h] * IC[h]) Where: • EC[x] is the direct employment coefficient for each "x" type of RE ([s] solar, [w] wind, [h] hydro). This is understood as the number of people needed to build 1 megawatt (MW) of installed capacity in each type of RE. • IC[x] is the new installed capacity building demand of each "x" type of RE. This comes from Georgia's CSAP (see activity scope).
Data sources	• Employment coefficient for building solar power plant based on international literature (2.5/MW) • Employment coefficient for building wind power plant based on public reports from Karti WPP (1.6/MW) • Employment coefficient for building hydropower plant based on international literature – 3.7/MW • Number of years to build solar power plant based on international literature – 1 year • Number of years to build wind power plant based on international literature – 2 years • Number of years to build hydropower plant based on international literature – 2 years • RE building scenario provided by government counterparts

CSAP = 2030 National Climate Strategy and 2021–2023 Action Plan, MW = megawatt.
Source: Authors.

Table A1.3: Job Creation for the Operation and Maintenance of Renewable Energy Projects

Description	Number of direct jobs created during the operation and maintenance (O&M) of the renewable energy (RE) power plants.
Quantification formula	Total Jobs = (EC[s] * IC[s])+ (EC[w] * IC[w])+ (EC[h] * IC[h]) Where: • EC[x] is the direct employment coefficient for each "x" type of RE ([s] solar, [w] wind, [h] hydro). This is understood as the number of people needed to operate and maintain 1 MW of installed capacity in each type of RE. • IC[x] is the new installed capacity built for each "x" type of RE.
Data sources	• Direct employment coefficient for the O&M of solar power plants based on public national registries ppp.gov.ge (1.9/MW) • Direct employment coefficient for the O&M of wind power plant based on public national registries ppp.gov.ge (0.5/MW) • Direct employment coefficient for the O&M of hydropower plant based on public national registries ppp.gov.ge (0.5/MW) • Average number of operating years for each type of RE power plant based on public national registries ppp.gov.ge (10+ years)

MW = megawatt.
Source: Authors.

Table A1.4: Job Opportunities for Local Communities Versus Unemployment

Description	**Local communities can benefit from new job opportunities thanks to the introduction of these RE projects.** To assess the importance and significance of these job opportunities, it is essential to consider the socioeconomic context of the locations where the RE projects would take place, in particular the number of unemployed people who could potentially benefit from these job opportunities. Job opportunities include: Temporary job opportunities_x = Σ Construction jobs created_i / Σ unemployed_i Permanent job opportunities_x= Σ O&M jobs created_i / Σ unemployed_i Where "x" represents the locations where the RE projects would take place.
Quantification formula	(1) Construction jobs created_i = EC_x*IC_i Where: • EC represents the employment coefficient needed for building 1 MW of installed capacity for each type of RE "x" (wind, solar, hydro), as estimated in variable # 1 • IC represents the installed capacity of each project "i" as defined in the activity scope. (2) O&M jobs created_i = EC_x*IC_i Where: • EC represents the employment coefficient needed for O&M of 1 MW of installed capacity for each type of RE "x" (wind, solar, hydro), as estimated in variable # 1 • IC represents the installed capacity of each project "i" as defined in the activity scope.
Data sources	• Number of unemployed people in each RE project location: Geostat • Number of unemployed: Geostat • Jobs created: Authors' calculations from previous variables

CSAP = 2030 National Climate Strategy and 2021–2023 Action Plan, MW = megawatt, O&M = operation and maintenance.
Source: Authors.

Table A1.5: Housing and Property Values of Local Communities

Description	The construction of renewable energy (RE) power plants **can raise several issues for local communities which, if not properly addressed, can make the implementation of these projects difficult.** These include a wide range of issues such as land use rights, forced migration, noise pollution, health and environmental issues, landscape aesthetics, cultural disruption, etc. These issues are tricky to quantify in a socioeconomic model as many of them are subjective and/or case-specific and **would need a project-level assessment**. **For the modeling of this assessment,** it was therefore decided to consider the **change in housing and property value as a global indicator to reflect these issues.** While the property value is not perfect, it should **reflect exogenous environmental, economic, and social factors that may have a direct or indirect impact on property.** ΔPropertyValues_x = Σ New Property values_x_i – Σ Initial Propery Values_x_i Where "i" represents the locations where the RE projects would take place and "x" represents the RE project types.
Quantification formula	(1) Initial property values_x= [Average Property Value_x_i * Number of Properties_x_i]i (2) New Property values_x = 1+ Δ Property Value_x * Initial Property value_x Where: "i" represents the RE project locations and "x" the RE type of project (solar, wind or hydro) Δ % Property Value_x represents the average change in property value (as a percentage) for each type of RE project "x"
Data sources	Average property value in each RE project location: data scaping from SS.ge

continued on next page

Table A1.5 *continued*

Assumptions	• Average impact on property value from solar power plant installation: New research from the Lawrence Berkeley National Laboratory has found that solar farms reduce property values of homes within a half-mile radius by an average of 1.5%. The study found no effect on homes 1 mile away.
	• Average impact on property value from wind power plant installation: Large windfarms can reduce the value of homes within a 2 kilometer (km) radius by up to 12% and reduce property prices up to 14 km away, according to research by the London School of Economics.
	• Average impact on property value from hydropower plant installation: 7%, based on various international sources.

Source: Authors.

Table A1.6: Government Tax Revenue from New Renewable Energy Projects

Description	Tax revenues are generated through taxes imposed on various aspects of the renewable energy (RE) project, such as corporate income tax, property tax, and value-added tax (VAT). Total Tax Revenues = Σ (CTr + PTr + VATr)$_{[x]}$ Where: • CTr is the corporate income tax revenue for the government from the RE project of RE type [x] • PTr is the property tax revenue for the government from the RE project of RE type [x] • VATr is the value-added tax revenue for the government from the RE project of RE type [x]
Quantification formula	Total Tax Revenues = [(CT * Profit) + (PT * PropertyValue) + (VAT*SalesValue)] [s] + [(CT * Profit) + (PT * PropertyValue) + (VAT*SalesValue)] $_{[w]}$ + [(CT * Profit) + (PT * PropertyValue) + (VAT*SalesValue)]$_{[h]}$ Where: • "Profit" is the average profit generated by each type of renewable energy project ([s] solar, [w] wind, [h] hydro). • "PropertyValue" is the average value of the property or assets associated with each type of RE project. • "SalesValue" is the average total value of sales of goods and services related to each type of RE project.
Data sources	• Average electricity tariffs for wind, solar and hydropower: Primary data provided by government counterparts: \$68.5/MW for hydropower, \$68.2/MWh for wind power, and \$63/MWh for solar power
	• Profit margin: Based on international literature, 10% profit margin considered from the energy tariff (*The Guardian*)
	• Energy generation capacity, based on international literature: 25% for solar, 40% for wind, 45% for hydropower
	• Property value: primary data provided by the government based on total investment
	• Tax rates: national public sources: 18% VAT, 1% property tax, 15% corporate income tax

MWh = megawatt-hour.
Source: Authors.

Table A1.7: Government Nontax Revenue from New Renewable Energy Projects

Description	Nontax revenues may include revenues from fees and permits (charges for licenses, permits, environmental assessments, and other regulatory requirements) and other contributions such as royalties and lease payments for land resources used. Total Nontax Revenues = Σ (FPr + Rr+ LPr)$_{[x]}$ Where: ● FPr is the revenue from fees and permits from the RE project of RE type [x] ● Rr are the other contribution revenues from the RE project of RE type [x]
Quantification formula	Total Nontax Revenues = [(FP* IC) + (R*IC) + (LP*IC)] $_{[s]}$ +[(FP* IC) + (R*IC) + (LP*IC)] $_{[w]}$ + [(FP* IC) + (R*IC) + (LP*IC)] $_{[h]}$ Where: ● FP are the fees and permits paid in average per MW of installed capacity for each type of RE project ● R are the royalties paid in average per MW of installed capacity for each type of RE project ● LP are the lease payments paid in average per MW of installed capacity for each type of RE project
Data sources	Average amount of fees and permits paid per MW of installed capacity for each type of RE project: reporting portal https://reportal.ge/en/Reports Solar: $700/MW Wind: $771.9/MW Hydropower: $1,530/MW ● Amount of royalties paid in average per MW of installed capacity for each type of RE project: Ministry of Economy: According to MOESD representatives, RE plants don't pay royalties. Amount of lease payments paid in average per MW of installed capacity for each type of RE project: Reporting Portal https://reportal.ge/en/Reports Solar: $70/MW Wind: $72/MW Hydropower: $209/MW

MOESD = Ministry of Economy and Sustainable Development, MWh = megawatt-hour, RE = renewable energy.
Source: Authors.

Table A1.8: Government Tax Revenues from New Employees in the Economy

Description	Another source of government revenue would be the taxes collected from the people who would become employed thanks to these projects. This includes tax revenue from jobs in the construction of RE plants and tax revenue from jobs in the O&M of RE plants. Total Government Tax Revenue = Tax Revenue from RE Construction Jobs + Tax Revenue from O&M jobs
Quantification formula	(1) Tax revenue from construction jobs = [(AIs * Ns*Ds*Tax_s) $_{[s]}$ + (AIw * Nw*Ds*Tax_w) $_{[w]}$ + (AIh*Nh*Dh*Tax_h) $_{[h]}$] Where: • AI is the average monthly income for each type of RE ([s] solar, [w] wind, [h] hydro). • N is the number of jobs created for the construction of each type of RE. • Ds is the average number of months to build each type of RE. • Tax_r is the average monthly income tax rate for each type of RE (based on average income for each RE). (2) Tax revenue from O&M jobs = [(AIs * Ns*Tax_s) $_{[s]}$ + (AIw * Nw*Tax_w) $_{[w]}$ + (AIh*Nh*Tax_h) $_{[h]}$] Where: • AI is the average annual income for each type of RE ([s] solar, [w] wind, [h] hydro). • N is the number of jobs created for the O&M of each type of RE. • Tax_r is the average annual income tax rate for each type of RE (based on average income for each RE).
Data sources	• Average monthly income for the construction jobs of each type of RE. Based on the average salary for a technician-level employee in Georgia ($12,000/year). World Salaries. Average Maintenance Mechanic Salary in Tbilisi, Georgia for 2024. https://worldsalaries.com/average-maintenance-mechanic-salary-in-tbilisi/georgia/ • Tax rate: National public data

MOESD = Ministry of Economy and Sustainable Development, MWh = megawatt-hour, O&M = operation and maintenance, RE = renewable energy.
Source: Authors.

(b)   Energy Efficiency Reform for the Construction of New Buildings

Table A1.9: Impact on Micro, Small and Medium-Sized Enterprises Suppliers

Description	This variable is used to determine the share of the potential increase in costs faced by local suppliers of materials and parts compared to the current total costs. This includes material costs, regulatory costs and training costs. The cost analysis should be performed accounting for the different sizes and revenues of suppliers. Indeed, larger suppliers may have more resources to absorb material, regulatory and training costs, while smaller suppliers may experience a greater proportional impact on their operations. Including these distributional impacts helps to capture these disparities and provide a more nuanced assessment.
Quantification formula	Cost for Local Suppliers_i = Σ (Δ MaterialCost_i + Δ RegulatoryCost_i + Δ TrainingCost_i) Where "i" represents each supplier category (by size and revenue). (1) ΔMaterialCost_i = InitialMaterialCost_i * ΔMaterialCost (2) ΔRegulatoryCost_i = InitialRegulatoryCost_i * ΔRegulatoryCost (3) ΔTrainingCost_i = (InitialTrainingCost_i * ΔtrainingCost Where the deltas (Δ) represent proportional changes (e.g., a 10% increase would be represented as 1.10.

continued on next page

Table A1.9 *continued*

Data sources	• Categorization of suppliers by size and revenue: Geostat Business Registry • Average costs for MSMEs suppliers (large and medium-sized): Reportal • Costs of machinery based on retail prices of the machines listed • Average regulatory costs based on international literature (International Energy Agency)
Assumptions	• Average costs for MSMEs: N/A for small and micro suppliers, therefore assumption based on average costs for medium-sized suppliers, related to the number of employees. The costs for micro and small enterprises were estimated based on the costs for medium-sized suppliers as a proportion of the median number of employees. • Increase in material costs: +20%

MSMEs = micro, small and medium-sized enterprises, N/A = not applicable.
Source: Authors.

Table A1.10: Distributional Impact on House Acquisition Prices

Description	This variable aims to analyze the potential impact of the increase in house acquisition prices on different household income groups to determine the effects on house acquisition affordability for various segments of the population.
Quantification formula	(1) ΔHouseAcquistionPrices_i= [HouseAcquisitionCost_i * (1+PriceIncrease%)] – HouseAcquisitionCost_i Where: • "i" represents each household income group • HouseAcquisitionCost_i represents the average cost of buying a house for each income group "i". • PriceIncrease% represents the average percentage increase in house acquisition prices. In the end, an alternative based on available data was provided.
Data sources	• Household income groups and annual revenue of each: National survey • Maximum share of income allowed for mortgage loans: Global Property Guide • Current share of income allocated: National household surveys • Gender wage gap provided by government counterparts (30%)

Source: Authors.

Table A1.11: Distributional Impact on Rent Expenditure

Description	This variable aims to analyze the potential impact of the increase in rental prices on different household income groups to determine the effects on the renting affordability for various segments of the population. Δ Share house renting prices_i = Δ ΔHouseRentingPrices_i/AnnualRevenue_i Where "i" represents each household income group.
Quantification formula	(1) Δ HouseRentingPrices_i= [HouseRentingCost_i * (1+PriceIncrease%)] – HouseRentingCost_i Where: • HouseRentingCost_i represents the average cost of renting a house for each income group "i". • Price Increase% represents the average percentage increase in house renting prices.
Data sources	• Household income groups and annual revenue of each: National survey • Average house rental spending for each income group: National survey
Assumptions	In a simplified scenario, it is assumed that if buildings become 5% more expensive due to energy efficiency measures, rent prices would follow a linear increase (5%).

Source: Authors.

Table A1.12: Distributional Impact on Energy Expenses

Description	This variable aims to analyze the potential impact of the change in electricity bills on different household income groups to determine the effects of the reform on electricity affordability for households that would acquire or rent these type of buildings compared to those that would not have this option. $\Delta ElectricityAffordability\_i = EE\_ExpWeight\_i - NonEE\_ExpWeight\_i$
Quantification formula	(1) Change in electricity consumption: ● $NewElecConsumption\_i = InitialElectricityConsumption\_i * DecreaseElectricityConsumption\%$ Where: ● $InitialElectricityConsumption\_i$ represents the average electricity consumption for households of each income group "i" ● $DecreaseElectricityConsumption\%$ represents the average percentage decrease in electricity consumption thanks to the reform (see assumption below). (2) Energy expenditure for each income group "i" in energy efficiency buildings: ● $NewExpenditure\_i = Price\_i * NewElecConsumption\_i$ (3) Share of electricity consumption in energy efficiency building relative to annual revenue: ● $EE\_ExpShare\_i = NewExpenditure\_i / AnnualRevenue\_i$ (4) Share of electricity consumption in non-energy efficiency building relative to annual revenue: ● $NonEE\_ExpShare\_i = InitialElectricityConsumption\_i / AnnualRevenue\_i$
Data sources	● Electricity consumption: National household survey ● Average income: National household survey ● Electricity price: EnergoPro Energy
Assumptions	● In a simplified scenario, it is assumed that if the buildings are 5% more expensive due to energy efficiency measures, households can save up to 20% in energy savings. However, this is a general estimate and the actual results can vary greatly depending on the specific circumstances and the measures implemented. It is essential to conduct a detailed energy audit or analysis to accurately assess the potential saving.

Source: Authors.

(c)   Intercity Bus Transport Reform

Table A1.13: Distributional Impact on Service Users

Description	This variable aims to analyze the potential impact of the increase in ticket prices on the different service users to determine the effects on service costs for the affected population.
Quantification formula	Impact on users demand was estimated based on the assumption of –0.56 elasticity. (1) Scenario 1: Price increase of 5% (2) Scenario 2: Price increase of 10% (3) Scenario 3: Price increase of 15%
Data sources	● Regulatory impact assessment ● Literature review on demand elasticity estimates
Assumptions	● Demand elasticity for bus transport is –0.56 ● Insurance costs are $100 ● Cost of acquiring a new vehicle is $39,070 ● The proof of financial status is not part of the upfront costs due to introduced proof alternatives considered by the government.

Source: Authors.

Table A1.14: Impact on Compliance Costs for Service Providers

Description	This variable aims to quantify the upfront costs that service providers will face when the reform is implemented. It includes one-time costs and periodic costs incurred by service providers. According to our information, most service providers are individual bus owners who also drive their vehicle. The calculations presented here are based on the cost this individual will have to bear.

Quantification formula	(1) Scenario of baseline compliance
	    Total costs S1 = Permit Cost + Station Fee + Insurance + Estimated Transport Manager Cost
	(2) Scenario of upgrade compliance
	    Total costs S2 = Total costs S1 + Bus acquisition + Tax(y)

Data sources	• Source: Explanatory Note • Local consultations • Market prices for vehicle and insurance-related costs
Assumptions	• Insurance costs are $100 • Hiring a transport manager costs $1,984, the total estimate is divided between four individual drivers. • Income tax rate is 1%, when applicable • Cost of acquiring a new vehicle is $39,070 • The proof of financial status is not part of the upfront costs due to introduced proof alternatives considered by the government.

Source: Authors.

Table A1.15: Impact on Local Communities—Workers in the Vicinity of the Bus Station

Description	This variable aims to measure the impact on vendors in the vicinity of the bus station and is based on the activity of the bus station
Quantification formula	Total impacted workers = Total workers relying on the station * Number of declared stations
Data sources	• Local consultations
Assumptions	• All stations have an average of 10 informal workers relying on the station activity • The same informal job activities occur, regardless of the station class.

Source: Authors.

Appendix 2. Assessment of Impact Significance

(a)   Construction of Renewable Energy Power Plants

Table A2.1: Impacts of Renewable Energy Power Plants Construction

Variable	Type	Severity	Likelihood	Justification	Significance
Variable 1: Distributional impact on electricity expenditure in household income groups	RISK	Low	Likely	**Low severity:** It may only have moderate consequences for a specific small group of the population. **Likely:** According to the international literature, it is very likely to happen, although some specificities related to the Georgian context lower the likelihood to medium.	MINOR
Variables 2–4: Job creation for the new RE power plants	OPPORTUNITY	Medium	Likely	**Medium severity:** There could be moderate consequences only for a specific small group of the population. **Likely:** Job opportunities will arise. The level of measures implemented by the government will influence the extent to which this opportunity benefits the local population.	MODERATE
Variable 5: Impact on property value and local communities	RISK	High	Likely	**High severity:** Renewable energy projects in Georgia face great social resistance. If this issue is not properly addressed, it may jeopardize the implementation of the project. **Likely:** The impact is likely to happen, especially for hydropower plants, based on past examples, although it can be prevented if the proper measures are taken.	MAJOR
Variables 6–9: Fiscal revenues	OPPORTUNITY	Medium	High	**Medium severity:** The projects will generate a significant source of revenue for the government, which can be used to implement just transition measures. High likelihood as the projects are already planned and the energy demand already exists.	MODERATE

RE = renewable energy.
Source: Authors.

(b)   Energy Efficiency Reform for the Construction of New Buildings

Table A2.2: Impacts of Energy Efficiency Reform for New Building Construction

Variable	Type	Severity	Likelihood	Justification	Significance
Variable 1: Distributional impact on constructors' suppliers	RISK	Medium	High	**Medium severity:** It may only have moderate consequences for a specific small group of the population. **High likelihood:** Due to the current level of development of MSMEs, it is highly likely to happen.	MODERATE
Variable 2: Distributional impact on constructors	RISK	Low	Unlikely	**Low severity:** Increase in costs will be passed on to the cost of the house. **Unlikely:** Due to the context of Georgia, it is unlikely that demand will decrease.	NEGLIGIBLE
Variable 3: Distributional impact on house acquisition affordability	RISK	Low	High	**Low severity:** Low consequences due to the Georgian housing context. **High/Inevitable:** The increase in price is highly likely to occur based on government estimations.	MINOR
Variable 4: Distributional impact on renting affordability	RISK	Low	High	**Low severity:** Low consequences due to the Georgian housing context. **High/Inevitable:** The increase in price is highly likely to occur based on government estimations.	MINOR
Variable 5: Distributional impact on energy savings	OPPORTUNITY	Medium	High	**Medium severity:** The reform could only bring moderate benefits for a noticeable segment of the population. High likelihood, as the reform is already implemented.	MODERATE

MSMEs = micro, small and medium-sized enterprises.
Source: Authors.

(c)   Intercity bus passenger transport reform

Table A2.3: Impacts of the Intercity Bus Passenger Reform

Variable	Type	Severity	Likelihood	Justification	Significance
Variable 1: Distributional impact on service users	RISK/ OPPORTUNITY	Medium	Likely	**Medium severity:** The impact may have moderate consequences, especially for low-income groups of the population. **Likely:** The impact is likely to happen, but it can be prevented if proper measures are taken.	MODERATE
Variable 2: Impact on compliance costs for service providers	RISK	Low	High	**Low severity:** It may have moderate consequences for a small group of the population. **High likelihood:** Due to cost of meeting the requirements of the reform.	MINOR
Variable 3: Impact on local communities – workers in the vicinity of bus stations	RISK	Low	Likely	**Low severity:** It may only have an impact on a small group living from the development of informal activities in the vicinity of the station. **Likely:** The impact is likely to happen, although it can be prevented if proper measures are taken.	MINOR

Source: Authors.

Appendix 3. Amendments to the Transport Laws

The amendments studied by the current socioeconomic impact assessment are as follows:

Table A3: Proposed Amendments to Existing Regulations

Amendments to the Law of Georgia "On Motor Transport"	**The definition for permitting activities** serves as the basis for the implementation of new types of permits.
	Increase carrier accountability toward passengers: Adapt passenger transportation to modern requirements. List of conditions and required documentation to be submitted.
	Rules and conditions for bus station certification together with specific requirements: Including the criteria for the term, issuance, refusal, suspension, renewal, and cancellation of bus station certificates.
Amendments to the Code of Administrative Offenses	Individuals found in violation are subject to an initial fine of GEL1,000.
	For repeated offenses, the fine increases to GEL2,000.
	The law stipulates that violations of the technical regulations for the operation of bus stations may result in fines of up to GEL5,000.
Amendments to the "Local Self-Government Code" in the Organic Law of Georgia	Strengthen the authority of municipalities by introducing new powers such as: (a) Making decisions regarding the suitability of establishing a bus station on municipality territory. (b) Specifying a list of settlements where it is not obligatory to board only at the bus station if there is no bus station within the administrative boundaries of these settlements.
Amendments to "Licenses and Permits"	**Include two new types of permits:** a permit for motor passenger transportation and a permit for international motor cargo transportation.
Amendments to "Control of Entrepreneurial Activity"	**The Land Transport Agency** can carry out inspections of business activities without the need for a court order.

Source: Authors.

Appendix 4. Summary of the Results of the Regulatory Impact Assessment

Table A4 summarizes the impacts mapped by the Regulatory Impact Assessment. Most of the indicators listed reflect the variables expected to be affected by the policy reform, and the columns show the impacts estimated by the qualitative/quantitative assessment.

Table A4: Results of the Regulatory Impact Assessment's Socioeconomic Impact Assessment

Type of Impact	Indicator	Expected Impact Alternative 1	Expected Impact Alternative 2	Expected Impact Alternative 3
Economic impact	Gross domestic product (GDP) created in the sector	Zero impact	Zero impact	Growth
	Employment	Unidentified	Unidentified	Growth
	Wage rates	Zero impact	Zero impact	Growth
	Investments in the sector	A slight increase	A slight increase	Significant growth
	Impact on ticket price	A slight increase	A slight increase	Growth
	Size of shadow economy (informal activities)	Decrease slightly	A significant reduction compared to the first scenario	Significant reduction
	Additional tax revenues from reducing informal sector	GEL5.3 million (current value) Depends on the quality of policy enforcement	GEL5.3 million (current value) Depends on the quality of policy enforcement	Significant growth
	Increase of tax revenues in the budget by improving accounting and control of revenues in the sector	NA	NA	Significant growth
Impact on the market	Quality of service	A slight improvement	A significant improvement over the first scenario	A significant improvement (depends on the enforcement process)
	Provision of services	It may decrease slightly in the short term, but no changes are expected in the long term.	It may decrease slightly in the short term, but no changes are expected in the long term.	A significant improvement
	Market concentration	The number of medium-sized and/or large companies in the market may increase slightly.	The number of medium-sized and/or large companies in the market may increase slightly.	Increasing the share of medium-sized and large companies in the sector.

continued on next page

Table A4 *continued*

Type of Impact	Indicator	Expected Impact Alternative 1	Expected Impact Alternative 2	Expected Impact Alternative 3
Policy enforcement costs	One-time cost of enforcement: the cost of implementing a risk assessment system and a national electronic registry	Increased by GEL120,200–GEL825,200 (depending on the complexity of the system	Increased by GEL120,200–GEL825,200 (depending on the complexity of the system)	Increased by GEL120,200–GEL825,200 (depending on the complexity of the system)
	Cost of compliance checks and inspections to control permit conditions	Growth (depends on inspection methods and inspection frequency)	Increase (depends on inspection methods and inspection frequency) additional GEL951,300 expenses for roadside inspections	Increase (depends on inspection methods and inspection frequency) additional GEL951,300 expenses for roadside inspections
	Cost of hiring additional staff to handle customer complaints on the hotline	NA	GEL514,000, with current value (present values) in 2022–2031	GEL514,000, with current value (present values) in 2022–2031
	Cost of transport plans, tender conditions, and tender organization	NA	NA	Growth
Policy compliance costs	One-time compliance costs for infrastructure and internal system communications	Growth	Growth	Growth
	Business administrative expense	Significant growth	Significant growth	Significant growth
	Compliance costs incurred by bus stations and service providers	An increase from GEL22.8 million to GEL133.1 million	An increase from GEL22.8 million to GEL133.1 million	An increase from GEL22.8 million to GEL133.1 million
	Cost of installation of tachographs and speed limiters	NA	Increase to GEL54.7 million	Increase to GEL54.7 million
Passenger safety	Passenger safety	Compliance with minimum safety standards defined by technical regulations will be slightly improved	Significant improvement in compliance with safety norms defined by technical regulations	Significant improvement in compliance with safety norms defined by technical regulations
	Observance of road safety norms by drivers	Zero impact	Improvement	Improvement
The social benefits of reducing traffic accidents	Reduced social costs by decreasing road deaths and injuries	NA	GEL53.8 million savings	GEL53.8 million savings

continued on next page

Table A4 *continued*

Type of Impact	Indicator	Expected Impact Alternative 1	Expected Impact Alternative 2	Expected Impact Alternative 3
Ecological consequences	Transport emissions (from route optimization)	Zero impact	Zero impact	GEL31. 3 million
	Transport emissions (from improving road worthiness testing)	Zero impact	Zero impact	Reduction
	Reduction of emissions as a result of renewal of the car fleet	NA	NA	GEL4.4 million

NA = not applicable.
Source: PMO Business Consulting. 2021. Assessing the Impact of Regulation to Strengthening Urban Connectivity in Georgia.

Appendix 5. Pre-Assessed Transport Impacts

Regarding the package of proposed changes in the law, the main impacts identified in the **Government's Explanatory Note** are listed in Table A5.

Table A5: Expected Impacts Mapped by the Explanatory Note

	Expected Impact	Cost Burden	Estimated Amount (GEL)	Remarks
INCOME				
State revenue	New payment fee From 1 May 2025: New payment fee = GEL100 After 1 May 2025: Fee = GEL500	Paid for road passenger transport permit (bus owners, which are in general the same as drivers)	For first year: Fee = (GEL4,125 + GEL4,500)*500 = **GEL4,312,500** For the next 3 years: Fee = 100*500 = GEL500,000 per year	Estimated in the Explanatory Note
	Formalization of irregular activity – "Predators"	Predators who are entitled with the permit	GEL5.3 million in 2022–2023	Estimated in the Regulatory Impact Assessment (RIA)
	Other taxes affected by the formalization of the activity	Service providers • Service users and providers	• Increase in profit and income tax • Increase in value-added tax	Not estimated in previous assessments
Land Transport Agency (LTA) income	Payment of fee (for drivers)	Depending on the number of fees		Not estimated in previous assessments
	Permits to bus stations	Depending on the number of undeclared bus stations that will be regulated through their certification		Not estimated in previous assessments
Municipalities	Other taxes affected by the formalization of the activity	• Bus station • Service users and providers	• Property tax (municipal budget) • Value-added tax	Not estimated in previous assessments

continued on next page

Table A5 *continued*

	Expected Impact	Cost Burden	Estimated Amount (GEL)	Remarks
STATE EXPENDITURES				
State budget	NA	Not affected		As per Explanatory Note
Land Transport Agency expenditures	New administrative expenses to finance measures: Printing permits and certificates, conducting the certification process of bus stations, arranging the appropriate material and technical base, adding and retraining human resources, inspecting market participants (monitoring and controlling the enforcement of permit conditions), as well as the so-called expenditure of policy implementation costs (preparing drafts, technical regulations, etc.)	LTA		Not estimated in previous assessments
	Monitoring and control of permit conditions in the sector	LTA	Depends on the number and frequency of inspections carried out annually	
	Cost of creating a national electronic registry	LTA	**Ranges from GEL96,300 to GEL665,500,** depending on the size and complexity of the system	Estimated by the RIA
	One-time costs related to the operation of the system	LTA	Ranges from GEL24,600 to GEL159,700	Estimated by the RIA
EXPECTED FINANCIAL CONSEQUENCES				
Law receptors	Payment of permit fee	Bus owners	Permit fee for a Georgian 5 year permit = **GEL500** (one-time) Permit fee for a foreign carrier = **GEL1,500** (one-time permit)	Estimated by the RIA

continued on next page

Table A5 *continued*

	Expected Impact	Cost Burden	Estimated Amount (GEL)	Remarks
	Additional costs	Service providers (permit seekers)	• Obtain a criminal record = **GEL20** • Cost of hiring a manager = **GEL1,500** per month on average • Cost of insuring the bus against damage = **GEL9,000** on average • Obligation of financial resources = 0, in case of having own vehicle or a legal leasing	Estimated in the Explanatory Note. A transitional period will be created for those who do not meet the requirements. The government has high expectations that all operators will be able to continue their activities.
Bus station fee	A cap is set on the fee = **10%** *(60%)*(total number of bus seats*ticket price)	Paid by the carriers, affecting the bus station "owners"	Estimated that the fee will be within 6%–10% of the revenue received by the service provider	Estimated in the Explanatory Note
Total revenues for the bus station owners				Not estimated in previous assessments
To the owners entitled with permits	An increase in revenues is expected			Not estimated in previous assessments. Depends on the number of buses using the station
Bus station carriers' additional costs	Increase in compliance costs for the following: • upgrading the infrastructure • purchasing the necessary equipment and techniques • adjusting internal organizational processes to new requirements • hiring qualified staff • retraining existing staff to meet the necessary qualification requirements • generally ensuring compliance with the conditions established by law	Paid by the bus station carriers		Not estimated in previous assessments

Source: Government of Georgia, Ministry of Economy and Sustainable Development. 2019. Explanatory Note on the draft Amendments to the Law of Georgia "On Motor Transport."

www.ingramcontent.com/pod-product-compliance
Lightning Source LLC
LaVergne TN
LVHW071449180726
843512LV00018B/1327